FOUR-DIMEN
VISTAS

FOUR-DIMENSIONAL VISTAS

CLAUDE BRAGDON

NEW YORK

Four-Dimensional Vistas

For information, address:

Cosimo
P.O. Box 416
Old Chelsea Station
New York, NY 10113-0416

or visit our website at:
www.cosimobooks.com

Four-Dimensional Vistas originally published by Alfred A. Knopf in 1941.

Library of Congress Cataloging-in-Publication Data
A catalog record for this book is available from the Library of Congress

Cover design by www.wiselephant.com

ISBN: 1-59605-398-4

"Perception has a destiny."

EMERSON.

PREFACE TO THE SECOND EDITION

Since the first publication of *Four-Dimensional Vistas* in 1916, there has been a marked access of interest in the subject with which it deals. This is partly because of the extraordinary popular interest in the Einstein Theory of Relativity, which in one of its aspects involves the idea of a fourth dimension; and partly by reason of Ouspensky's *Tertium Organum*, a widely circulated book which deals with these same ideas philosophically, with a competence to which *Vistas* can lay no claim.

Inevitably my own point of view has been modified by Ouspensky's, but not to the extent of making it necessary to recast *Vistas*, which will be found to be complementary and not in any way contradictory to *Tertium Organum*. Ouspensky himself recognized this, for after reading *Vistas* he wrote me that he was "startled" at the parallelism between his thought and mine. It is as though two persons were contemplating the same metaphysical "scene" from different points of view and through different temperaments—one of them an amateur and an artist, and the other a logician and philosopher.

This similarity of content is one of the grounds

(but not the only one) for our common belief that some "higher" consciousness operates in us, and also in many others. We believe that men of one "mind" are being enlightened, prepared and united for the purpose of bringing about a recrudescence of mysticism, by providing it with a mathematical and philosophical foundation which it has appeared to lack.

CLAUDE BRAGDON.

NEW YORK, *April 1922.*

INTRODUCTION

There are two notable emancipations of the mind from the tyranny of mere appearances that have received scant attention save from mathematicians and theoretical physicists.

In 1823 Bolyai declared with regard to Euclid's so-called axiom of parallels, "I will draw two lines through a given point, both of which will be parallel to a given line." The drawing of these lines led to the concept of the curvature of space, and this to the idea of *higher* space.

The recently developed Theory of Relativity has compelled the revision of the time-concept as used in classical physics. One result of this has been to introduce the concept of *curved* time.

These two ideas, of curved time and higher space, by their very nature are bound to modify human thought profoundly. They loosen the bonds within which advancing knowledge has increasingly labored, they lighten the dark abysses of consciousness, they reconcile the discoveries of Western workers with the inspirations of Eastern dreamers; but best of all, they open vistas, they offer "glimpses that may make us less forlorn."

FOUR-DIMENSIONAL VISTAS

CONTENTS

Contents

I

THE QUEST OF FREEDOM

THE UNDISCOVERED COUNTRY

EXPECTANCY of freedom is the dominant note of today. Amid the clash of armies and the crash of systems we await some liberating stroke which shall release us from the old dreary thraldoms. As Nietzche says, "It would seem as though we had before us, as a reward for all our toil, a country still undiscovered, the horizons of which no one has yet seen, a beyond to every country and every refuge of the ideal that man has ever known, a world so overflowing with beauty, strangeness, doubt, terror and divinity, that both our curiosity and our lust of possession are frantic with eagerness."

Should a name be demanded for this home of freedom, there are those who would unhesitatingly call it the *Fourth Dimension of Space.* For such readers as may be ignorant of the amazing content of this seemingly meaningless phrase, any summary attempt at enlightenment will lead only to deeper mystification. To the question, where and what is the fourth dimension, the answer must be: it is here—in us and all about us—in a direction toward which we can never point because at right angles to every direction that we know. Our space cannot contain it, because it contains our space. No walls separate us from this demesne, not even the walls of our fleshly prison; yet we may not enter, even though we are already "there." It is the place of dreams, of living dead men; it is *At the Back of the North Wind* and *Behind the Looking Glass.*

So might one go on, piling figure upon figure and paradox upon paradox, to little profit. The effective method is the ordered and deliberate one; therefore the author asks of his reader the endurance of his curiosity pending certain necessary preparations of the mind.

MIRACLES

Could one of our aviators have landed in ancient Athens, doubtless he would have been given a place in the Greek Pantheon, for the old idea of a demigod was a man with wings. Why then does a flying man so little amaze us? Because we know about engines, and the smell of gasoline has dulled our sense of the sublime. The living voice of a dead man leaves us unterrified if only we can be sure that it comes from a phonograph; but let that voice speak to us out of vacancy and we fall a prey to the same order of alarm that is felt by a savage at the report of a gun that he has never seen.

This illustration admirably defines the nature of a miracle: it is a manifestation of power new to experience, and counter to the current thought of the time. Miracles are therefore always in order, they always happen. It is nothing that the sober facts of today are more marvelous than the fictions of Baron Munchausen, so long as we understand them: it is everything that phenomena are multiplying that we are unable to understand. This increasing pressure upon consciousness *from a new direction* has created a need to found belief on something firmer than a bottomless gullibility of mind. This book is aimed to meet that need by giving the mind the freedom of new spaces; but before it can even begin to do so, the reader must be brought to see the fallacy of attempting to measure the limits of the possible by that faculty known as common sense—and by common sense is meant, not the appeal to abstract reason, but to concrete experience.

THE FAILURE OF COMMON SENSE

Common sense had scarce had its laugh at Bell, and its shout of "I told you so!" at poor Langley, when lo! the telephone became the world's nervous system, and aeroplanes began to multiply like summer flies. To common sense the alchemist's dream of transmuting lead into gold seems preposterous, yet in a hundred laboratories radium is breaking down into helium, and the new chemistry bids fair to turn the time-honoured jeer at the alchemists completely upside down. A wife whose mind was oriented in the new direction effectually silenced her husband's ridicule of what he called her credulity by reminding him that when wireless telegraphy was first suggested he had exclaimed, "Ah, that, you know, is one of the things that is not possible!" He was betrayed by his common sense.

The lessons such things teach us are summed up in the reply of Arago, the great savant, to the wife of Daguerre. She asked him if he thought her husband was losing his mind because he was trying to make permanent the image in a mirror. Arago is said to have answered, "He who, outside of pure mathematics, says a thing is impossible, speaks without reason."

Common sense neither leads nor lags, but is ever limited to the passing moment: the common knowledge of today was the mystery and enchantment of the day before yesterday, and will be the mere commonplace of the day after tomorrow. If common sense can so little anticipate the ordinary and orderly advancement of human knowledge, it is still less able to take that leap into the dark which is demanded of it now. The course of wisdom is therefore to place reliance upon reason and intuition, leaving to common sense the task of guiding the routine affairs of life, and guiding these alone.

THE FUNCTION OF SCIENCE

In enlisting the aid of reason in our quest for freedom, we shall be following in the footsteps of mathematicians and theoretical physicists. In their arduous and unflinching search after truth they have attained to a conception of the background of phenomena of far greater breadth and grandeur than that of the average religionist of today. As a mathematician once remarked to a neo-theosophist, "Your idea of the ether is a more material one than the materialist's own." Science however has imposed upon itself its own limitations, and in this connection these should be clearly understood.

Science is that knowledge which can be gained by exact observation and correct thinking. If science makes use of any methods but these it ceases to be itself. Science has therefore nothing to do with morals: it gives the suicide his pistol, the surgeon his life-saving lance, but neither admonishes nor judges them. It has nothing to do with emotion: it exposes the chemistry of a tear, the mechanism of laughter; but of sorrow and happiness it has naught to say. It has nothing to do with beauty: it traces the movements of the stars, and tells of their constitution; but the fact of their singing together, and that "such harmony is in immortal souls," it leaves to poet and philosopher. The timbre, loudness, pitch, of musical tones is a concern of science; but for this a Beethoven symphony is no better than the latest ragtime air from the music halls. In brief, science deals only with *phenomena*, and its gift to man is the power over his material environment.

MATHEMATICS

The gift of pure mathematics, on the other hand, is primarily to the mind and spirit: the fact that man uses it to get himself out of his physical predicaments is more or less by the way. Consider for a moment this paradox: mathematics, the very thing common sense swears by and dotes on, contradicts common sense at every turn. Common sense balks at the idea of *less than nothing;* yet the *minus* quantity, which in one sense is less than nothing, in that something must be added to it to make it equal to nothing, is a concept without which algebra would have to come to a full stop. Again, the science of quaternions, or more generally, a vector analysis, in which the progress of electrical science is essentially involved, embraces (explicitly or implicitly) the extensive use of *imaginary* or *impossible* quantities of the earlier algebraists. The very words "imaginary" and "impossible" are eloquent of the defeat of common sense in dealing with concepts with which it cannot practically dispense, for even the negative or imaginary solutions of imaginary quantities almost invariably have some physical significance. A similar statement might also be made with regard to *transcendental* functions.

Mathematics, then, opens up ever new horizons, and its achievements during the past one hundred years give to thought the very freedom it seeks. But if science is dispassionate, mathematics is even more austere and inhuman. It cares not for teeming worlds and hearts insurgent, so long as in the pure clarity of space *relationships* exist. Indeed, it requires neither time nor space, number nor quantity. As the mathematician approaches the limits already achieved by study, the colder and thinner becomes the air and the fewer the contacts with the affairs of every-day. The Promethean fire of pure mathematics is perhaps the greatest of all in man's catalogue of gifts; but it is not most itself, but least so, when, immersed in the manifoldness of phenomenal life, it is made to serve purely utilitarian ends.

INTUITION

Common sense, immersed in the mere business of living, knows no more about life than a fish knows about water. The play of reason upon phenomena dissects life, and translates it in terms of inertia. The pure logic of mathematics ignores life and disdains its limitations, leading away into cold free regions of its own. Now our desire for freedom is not to vibrate in a vacuum, but to live more abundantly. *Intuition* deals with life directly, and introduces us into life's own domain: it is related to reason as flame is related to heat. All of the great discoveries in science, all of the great solutions in mathematics, have been the result of a *flash* of intuition, after long brooding in the mind. *Intuition illumines.* Intuition is therefore the light which must guide us into that undiscovered country conceded by mathematics, questioned by science, denied by common sense—*The Fourth Dimension of Space.*

OUR SENSE OF SPACE

Space has been defined as "room to move about." Let us accord to this definition the utmost liberty of interpretation. Let us conceive of space, not alone as room to move ponderable bodies in, but as room to think, to feel, to strike out in unimaginable directions, to overtake felicities and knowledges unguessed by experience and preposterous to common sense. Space is not measurable: we attribute dimensionality to space because such is the method of the mind; and that dimensionality which we attribute to space is progressive, because progression is a law of the mind. The so-called dimensions of space are to space itself as the steps that a climber cuts in the face of a cliff are to the cliff itself: they are not necessary to the cliff, they are necessary only to the climber. Dimensionality is the mind's method of mounting to the idea of the infinity of space. When we speak of the fourth dimension, what we mean is the fourth stage in the apprehension of that infinity. We might as legitimately speak of a fifth dimension, but the profitlessness of any discussion of a fifth and higher stages lies in the fact that they can be intelligently approached only through the fourth, which is still largely unintelligible. The case is like that of a man, promised an increase of wages after he has worked a month, who asks for his second month's pay before he is entitled to the first.

THE SUBJECTIVITY OF SPACE

Without going deep into the doctrine of the ideality—that is, the purely subjective reality—of space, it is easy to show that we have arrived at our conception of a space of three dimensions by an intellectual process. The sphere of the senses is two-dimensional: except for the slight aid afforded by binocular vision, sight gives us moving pictures *on a plane*, and touch contacts *surfaces* only. What circumstances, we may ask, have compelled our intellect to conceive of *solid* space? This question has been answered in the following way:

"If a child contemplates his hand, he is conscious of its existence in a double manner—in the first place by its tangibility, and next by its image on the retina of his eye. By repeated groping about and touching, the child knows by experience that his hand retains the same form and extension through all the variations of distance and position under which it is observed, notwithstanding that the form and extension of the image on the retina constantly change with the different position and distance of his hand in respect to his eye. The problem is thus set to the child's understanding: how to reconcile to his comprehension the apparently contradictory facts of the *invariableness* of the object together with the *variableness* of its appearance. This is only possible within a space of three dimensions, in which, owing to perspective distortions and changes, these variations of projection can be reconciled with the constancy of the form of a body."

Thus we have come to the idea of a three-dimensional space in order to overcome the apparent contradictoriness of facts of sensible experience. Should we observe in three-dimensional space contradictory facts, our reason would be forced to reconcile these contradictions also, and if they could be reconciled by the idea of a four-dimensional space our reason would accept this idea without cavil. Furthermore, if from our childhood phenomena had been of daily occurrence requiring a space of four or more dimensions for an explanation conformable to reason, we should feel ourselves native to a space of four or more dimensions.

Poincaré, the great French mathematician and physicist, arrived at these same conclusions by another route. By a process of mathematical reasoning, of a sort too technical to be appropriately given here, he discovers an order in which our categories range themselves naturally, and which corresponds with the points of space; and that this order presents itself in the form of what he calls a "three circuit distribution board." "Thus the characteristic property of space," he says, "that of having three dimensions, is only a property of our distribution board, *a property residing, so to speak, in human intelligence.*" He concludes that a different association of ideas would result in a different distribution board, and that might be sufficient to endow space with a fourth dimension. He concludes that there may be thinking beings, living in our world, whose distribution board has four dimensions, and who consequently think in hyperspace.

THE NEED OF AN ENLARGED SPACE-CONCEPT

It is the contrariety in phenomena already referred to that is forcing advanced minds to entertain the idea of higher space. Mathematical physicists have found that experimental contradictions disappear if, instead of referring phenomena to a set of three space axes and one time axis of reference, they be referred to a set of four interchangeable axes involving four homogeneous co-ordinates. In other words, *time* is made the fourth dimension. Psychic phenomena indicate that occasionally, in some individuals, the will is capable of producing physical movements for whose geometrico-mathematical definition a four-dimensional system of co-ordinates is necessary. This is only another step along the road which the human mind has always travelled: our conception of the cosmos grows more complete and more just at the same time that it recedes more and more beneath the surface of appearances.

Far from the Higher Space Hypothesis complicating thought, it simplifies by synthesis and co-ordination in a manner analogous to that by which plane geometry is simplified when solid geometry becomes a subject of study. By immersing the mind in the idea of many dimensions we emancipate it from the idea of dimensionality. But the mind moves most readily, as has been said, in ordered sequence. Frankly submitting ourselves to this limitation, even while recognizing it as such, let us learn such lessons from it as we can, serving the illusions that master us until we have made them our slaves.

II

THE DIMENSIONAL LADDER

LEARNING TO THINK IN TERMS OF SPACES

THE reader who is willing to consider the Higher Space Hypothesis seriously, who would discover, by its aid, new and profound truths closely related to life and conduct, should first of all endeavor to arouse in himself a new Power of perception. This he will best accomplish by learning to discern dimensional sequences, not alone in geometry, but in the cosmos and in the natural world. By so doing he may erect for himself a veritable Jacob's ladder,

"Pitched between Heaven and Charing Cross."

He should accustom himself to ascend it, step by step, dimension by dimension. Then he will learn to trust Emerson's dictum, "Nature geometrizes," even in regions where the senses fail him, and the mind alone leads on. Much profitable amusement is to be gained by such exercises as follow. They are in the nature of a running up and down the scales in order to give strength and flexibility to a new set of mental fingers. Learning to think in terms of spaces contributes to our emancipation from the tyranny of space.

FROM THE COSMOS TO THE CORPUSCLE

By way of a beginning, proceed, by successive stages, from the contemplation of the greatest thing conceivable to the contemplation of the most minute, and note the space sequences revealed by this shifting of the point of view.

The greatest thing of which we can form any conception is the starry firmament made familiar to the mind through the study of astronomy. No limit to this vastitude has ever been assigned. Since the beginning of recorded time, the earth together with the other planets and the sun has been speeding through interstellar space at the rate of 300,000,000 miles a year without meeting or passing a single star. A ray of light, traveling with a velocity so great as to be scarcely measurable within the diameter of the earth's orbit, takes years to reach even the nearest star, centuries to reach those more distant. Viewed in relation to this universe of suns, our particular sun and all its satellites—of which the earth is one—shrinks to *a point* (a *physical* point, so to speak—not a geometrical one).

The mind recoils from these immensities: let us forsake them then for more familiar spaces, and consider the earth in its relation to the sun. Our planet appears as a *moving point,* tracing out a line —*a one-space*—its path around the sun. Now let us remove ourselves in imagination only far enough from the earth for human beings thereon to appear as minute moving things, in the semblance let us say of insects infesting an apple. It is clear that from this point of view these beings have a freedom of movement in their "space" (the surface of the earth), of which the larger unit is not possessed; for while the earth itself can follow only *a line,* its inhabitants are free to move in the two dimensions of *a plane*— the surface of the earth.

Abandoning our last coign of vantage, let us descend in imagination and mingle familiarly among men. We now perceive that these creatures which from a distance appeared as though flat upon the earth's surface, are really *solids* and that they are endowed with the power to move their members in *three dimensions.* Indeed, man's ability to traverse the surface of the earth is wholly dependent upon his power of three-dimensional movement. Observe that with each transfer of our attention from greater units to smaller, we appear to be dealing with a power of movement in an additional dimension.

Looking now in thought not *at* the body of man, but *within* it, we apprehend an ordered universe immensely vast in proportion to that physical ultimate we name the electron, as is the firmament immensely vast in proportion to a single star. It has been suggested that in the infinitely minute of organic bodies there is a power of movement in a *fourth* dimension. If so, such four-dimensional movement may be the proximate cause of the phenomenon of *growth*—of those chemical changes and renewals whereby an organism is enabled to expand in three-dimensional space, just as by a three-dimensional power of movement (the act of walking) man is able to traverse his two-dimensional space—the surface of the earth.

—AND BEYOND

Proceed still further: behind such organic change—assumed to be four-dimensional—there is the determination of some *will-to-live,* which manifests itself to consciousness as thought and as desire. Into these the idea of space does not enter: we think of them as in *time.* But if there are developments of other dimensions in space, thought and emotion may themselves be discovered to have space relations; that is, they may find expression in the forms of *higher* spaces. Thus is opened up one of those rich vistas in which the subject of the fourth dimension abounds, but into which we can only glance in passing. If there are such higher-dimensional *thought-forms,* our normal consciousness—which "works" only in three dimensions—can apprehend only their three-dimensional aspects, and these not simultaneously, but successively—that is, in *time.* According to this view, any unified series of *actions* —for example, the life of an individual, or of a group—would represent the straining, so to speak, of a thought-form through our *time,* as the bodies subject to these actions would represent its straining through our space.

EVOLUTION AS SPACE-CONQUEST

Evolution is a struggle for and a conquest of space; for evolution, as the word implies, is a *drawing out* of what is inherent, from latency into objective reality, or in other words into spatial—and temporal—extension.

This struggle for space, by means of which the birth and growth of organisms is achieved, is the very texture of life, the plot of every drama. Cells subdivide; micro-organisms war on one another; plants contend for soil, light, moisture; flowers cunningly suborn the bee to bring about their nuptials; animals wage deadly warfare in their rivalry to bring more hungry animals into a space-hungry world. Man is not exempt from this law of the jungle. Nations intrigue and fight for land—of which wealth is only the symbol—and a nation's puissance is measured by its power to push forward into the territory of its neighbour. The self-same impulse drives the individual. One measure of the difference between men in the matter of efficiency is the amount of space each can command: one has a house and grounds in some locality where every square inch has an appreciable value, another some fractional part of a lodging house in the slums. When this bloodless but none the less deadly contest for space becomes acute, as in the congested quarters of great cities, man's ingenuity is taxed to devise effective ways of augmenting his *space-potency*, and he expands in a vertical direction. This third-dimensional extension, typified in the tunnel and in the skyscraper, is but the latest phase of a conquest of space which began with the line of the pioneer's trail through an untracked wilderness.

DIMENSIONAL SEQUENCES

Not only does nature everywhere geometrize, but she does so in a particular way, in which we discover dimensional sequences. Consider the transformation of solid, liquid, gas, from one to another under the influence of heat. A solid, set in free motion, can follow only *a line*—as in the case of a thrown ball. A liquid has the added power of lateral extension. Its tendency when intercepted is to spread out in the two dimensions of *a plane*—as in the case of a griddle cake; while a gas expands *cubically* in all directions, as is shown by a soap-bubble. It is a reasonable inference that the fourth state of matter, the corpuscular, is affiliated to some four-dimensional manner of extension, and that there may be states beyond this involving an even higher development of space.

Next glance at the vegetable kingdom. The seed, *a point*, generates *a line* system—in stem, branches, twigs—from which depend *planes* in the form of leaves and flowers, and from these come fruit, *solids*.

> "The point, the line, the surface and the sphere,
> In seed, stem, leaf and fruit appear."

A similar sequence may be noted within the body: the *line*-network of the nerves conveys the message of sensation from *the surface* of the body to some centre in *the solid*, of the brain—and thence to the Silent Thinker, "He who is without and within," or in terms of our hypothesis, "He who dwells in higher space."

MAN THE GEOMETER

When man essays the rôle of creator he cannot do otherwise than follow the same sequence: it is easy to discern dimensional progression in the products of man's ingenuity and skill. Consider for example the evolution of a building from its inception to its completion. It exists first of all in the mind of the architect, and there it is indubitably higher-spatial, for he can interpenetrate and examine every part, and he can consider it all at once, viewing it simultaneously from without and from within, just as one would be able to do in a space of four dimensions. He begins to give his idea physical embodiment by making with a pencil-*point, lines* on *a plane* (a piece of paper), the third dimension being represented by means of the other two. Next (if he is careful and wise) he makes a model in three dimensions. From the architect's drawings the engineer establishes his points, lays out his angles, and runs his lines upon the site itself. The mason follows, and with his footing-courses makes ponderable and permanent the lines of the engineer. These lines become in due course walls—vertical planes; floors and roofs—horizontal planes—follow, until some portion of three-dimensional space has been enclosed.

Substantially the same sequence holds, whatever the kind of building or the character of the construction—whether a steel-framed skyscraper or a wooden shanty. A line system, represented by columns and girders in the one case, and by studs and rafters in the other, becomes by overlay or interposition a system of planes, so assembled and correlated as to define a solid.

With nearly everything of man's creating—be it a bureau or a battleship—the process is as above described: first a pattern to scale; next an actual linear framework; then planes defining a solid. Consider almost any of the industries practised throughout the ages, they may be conceived of thus in terms of dimensions: for example, those ancient ones of weaving and basket making. *Lines* (threads in the one case, rushes in the other) are wrought into *planes* to clothe a body or to contain a burden. Or think if you choose of the modern industry of book-making, wherein types are assembled, impressed upon sheets of paper, and these bound into volumes—*points, lines, planes, solids.* The book in turn becomes the unit of another dimensional order in the library, whose serried shelves form lines which combined into planes define the lateral limits of the room.

HIGHER—AND HIGHEST—SPACE

These are truisms. What have they to do, it may be asked, with the idea of *higher* spaces? They have everything to do with it, for in achieving the enclosure of any portion of solid space the limit of known dimensions has been reached without having come to any end. More dimensions—higher spaces—are required to account for higher things. All of the products of man's ingenuity are inanimate except as he himself animates them. They remain as they were made: machines, not organisms. They have no inherent life of their own, no power of growth and renewal. In this they differ from animate creation, because the highest achievement of the creative faculty in man in a mechanical way lacks the life principle possessed by the plant. And as the most perfect machine is inferior in this respect to the humblest flower that grows, so is the highest product of the vegetable kingdom inferior to man himself, the maker of the machine; for he can reflect upon his own and the world's becoming, while the plant can only become.

What is the reason for these differences of power and function? According to the Higher Space Hypothesis they are due to varying potencies of movement in the secret causeways and corridors of space. The higher functions of consciousness—volition, emotion, intellection—may be correlated in some way with the higher powers of numbers, and with the corresponding higher developments of space. Thus the difference between physics and metaphysics would become a difference of degree and not of kind. Evolution is to be conceived of as a continuous pushing back of the boundary between representation and reality, or as a conquest of space. We may conceive of space as of an infinite number of dimensions, and of consciousness as a moving—or rather as an expanding—point, embracing this infinity, involving worlds, powers, knowledges, felicities, within itself in everlasting progression.

III

PHYSICAL PHENOMENA

LOOKING FOR THE GREATER IN THE LESS

AFTER the assured way in which the author has conducted the reader repeatedly up and down the dimensional ladder, it may be a surprise to him to learn that physical phenomena offer no irrefragable evidences of hyper-dimensionality. We could not think in higher space if consciousness were limited to three dimensions. The mathematical reality of higher space is never in question: the higher dimensions are as valid as the lower; but the hyper-dimensionality of "matter" is still unproven—man's ant-like efforts to establish this as a truth thus far have been in vain.

Lest this statement discourage the reader at the very outset, he should understand the reason for such failure. We are *embedded* in our own space, and if that space be embedded in higher space, how are we going to discover it? If space be curved, how are we going to measure its curvature? Our efforts to do so may be compared to measuring the distance between the tips of a bent bow by measuring along the bow instead of along the string.

Imagine a scientifically-minded threadworm to inhabit a page of Euclid's solid geometry: the evidences of three-dimensionality are there, in the very diagrams underneath his eyes; but you could not *show* him a solid—the flat page could not contain it, any more than our space can contain a form of four dimensions. You could only say to him, "These lines *represent* a solid." He would have to depend on his *faith* and not on that "knowledge gained by exact observation and correct thinking" in which alone the scientist finds a sure ground for belief.

It is an axiom of science never to look outside three-space horizons for an understanding of phenomena when these can be accounted for logically within those horizons. Now because on the Higher Space Hypothesis each space is the container of all phenomena of its own order, the futility for practical purposes of going outside is at once apparent. The highly intelligent threadworm neither knows nor cares that the point of intersection of two lines in his diagram *represents* a point in a space to which he is a stranger. The point is there, on his page: it is what he calls *a fact.* "Why raise (he says) these puzzling and merely academic questions? Why attempt to turn the universe completely upside down?"

But though no *proofs* of hyper-dimensionality

have been found in nature, there are equally no contradictions of it, and by using a method not inductive but deductive, the Higher Space Hypothesis is in a measure confirmed. Nature affords a sufficient number of *representations* of four-dimensional forms and movements to justify their consideration here:

SYMMETRY

Let us first flash the light of our hypothesis upon an all but universal characteristic of living forms, yet one of the most inexplicable—*symmetry*.

Animal life exhibits the phenomenon of the right- and left-handed symmetry of solids. This is exemplified in the human body, wherein the parts are symmetrical with relation to the axial *plane*. Another more elementary type of symmetry is characteristic of the vegetable kingdom. A leaf in its general contour is symmetrical, the symmetry being about *a line*—the midrib. This type of symmetry is readily comprehensible, for it involves simply a revolution through 180 degrees. Write a word on a piece of paper and quickly fold it along the line of writing so that the wet ink repeats the pattern, and you have achieved the kind of symmetry represented in a leaf.

With the symmetry of solids, or symmetry with relation to an axial *plane*, no such simple movement as the foregoing suffices to produce or explain it, because symmetry about a plane implies *four-dimen-*

sional movement. It is easy to see why this must be so. In order to achieve symmetry in any space—that is, in any given number of dimensions—there must be revolution in the next higher space: one more dimension is necessary. To make the (two-dimensional) ink figure symmetrical, it had to be folded over *in the third dimension.* The revolution took place about the figure's *line* of symmetry, and in a *higher* dimension. In *three*-dimensional symmetry (the symmetry of solids) revolution must occur about the figure's *plane* of symmetry, and in a higher (i. e. in *a fourth*) dimension. Such a movement we can reason about with mathematical precision: we see the result in the right- and left-handed symmetry of solids, but we cannot picture the movement to ourselves because it involves a space of which our senses fail to give any account.

Now could it be shown that the two-dimensional symmetry observed in nature is the result of a three-dimensional movement, the right- and left-handed symmetry of solids would by analogy be the result of a *four*-dimensional movement. Such revolution (about a plane) would be easily achieved, natural and characteristic in four space, just as the analogous movement (about a line) is easy, natural, and characteristic in our space of three dimensions.

OTHER ALLIED PHENOMENA

In the mirror-image of a solid we have a representation of what would result from a four-dimensional revolution, the surface of the mirror being the plane about which the movement takes place. If such a change of position were effected in the constituent parts of a body as a mirror image of it *represents*, the body would have undergone a revolution in the fourth dimension. Now two varieties of tartaric acid crystallize in forms bearing the relation to one another of object to mirror-image. It would seem more reasonable to explain the existence of these two identical but reversed varieties of crystal, by assuming the revolution of a single variety in the fourth dimension, than by any other method.

There are two forms of sugar found in honey, dextrose and levulose. They are similar in chemical constitution, but the one is the reverse of the other when examined by polarized light—that is, they rotate the plane of polarization of a ray of light in opposite ways. If their atoms are conceived to have the power of motion in the fourth dimension, it would be easy to understand why they differ. Certain snails present the same characteristics as these two forms of sugar. Some are coiled to the right and others to the left; and it is remarkable that, like dextrose and levulose, their juices are optically the reverse of each other when studied by polarized light.

Revolution in the fourth dimension would explain the change in a body from producing a right-handed, to producing a left-handed, polarization of light.

ISOMERISM

In chemistry the molecules of a compound are assumed to consist of the atoms of the elements contained in the compound. These atoms are supposed to be at certain distances from one another. It sometimes happens that two compound substances differ in their chemical or physical properties, or both, even though they have like chemical elements in the same proportion. This phenomenon is called isomerism, and the generally accepted explanation is that the atoms in isomeric molecules are differently arranged, or grouped, in space. It is difficult to imagine how atoms, alike in number, nature, and relative proportion, can be so grouped as somehow to produce compounds with different properties, particularly as in three-dimensional space four is the greatest number of points whose mutual distances, six in number, are all equal and independent. In four-dimensional space, however, there is greater freedom of arrangement; there the *ten* equal distances between any two of *five* points are geometrically independent, thus greatly augmenting the number and variety of possible arrangements of atoms.

This just escapes being the kind of proof demanded by science. If the independence of all the possible distances between the atoms of a molecule is absolutely required by theoretical chemical research, then science is really compelled—in dealing with molecules of more than four atoms—to make use of the idea of a space of more than three dimensions.

THE ORBITAL MOTION OF SPHERES: CELL SUB-DIVISION

There is in nature another representation of hyper-dimensionality which though difficult to demonstrate is too interesting and significant to be omitted here.

Imagine a helix, intersected, in its vertical dimension, by a moving plane. If necessary to assist the mind, suspend a spiral spring above a pail of water, then raise the pail until the coils become immersed one after another. The spring would represent the helix, and the surface of the water the moving plane. Concentrating attention upon this surface you would see a point—the elliptical cross-section of the wire where it intersected the plane—moving round and round in a circle. Next conceive of the wire itself as a lesser helix of many convolutions, and repeat the experiment. The point of intersection would then continually return upon its own track in a series of minute loops forming those lesser loops which, moving circle-wise, registered the involvement of the helix in the plane.

It is easy to go on imagining complicated structures of the nature of the spiral, and to suppose also that these structures are distinguishable from each other at every section. If we think of the intersection of these with the rising surface as the "circular" atoms, or physical units, of a plane universe, we shall have a world of apparent motion, with bodies moving harmoniously amongst one another, each a cross-section of some part of an unchanging and unmoving three-dimensional entity.

Now augment the whole by an additional dimension—raise everything one space. The helix of many helices would become four-dimensional, and superficial space would change to solid space: each tiny *circle* of intersection would become *a sphere* of the same diameter, describing, instead of loops, helices. Here we would be among familiar forms, describing familiar motions: the forms for example of the earth and the moon and their movement about the sun; of the atom (as we imagine it), the molecule and the cell. For is not the sphere, or ovoid, the unit form of nature; and is not the spiral vortex its characteristic motion, from that of the nebula in the sky to the electron in the atom? Thus on the hypothesis that our space is traversing four-dimensional space, and that the forms of our space are cross-sections of four-dimensional forms, the unity and harmony of nature would be accounted for in a remarkably simple manner.

The above exercise of the imagination is a good preparation for the next demand upon it. Conceive a dichotomous tree—one that always divides into two branches—to pass through a plane. We should have, as a plane section, a circle of changing size, which would elongate and divide into two circles, each of which would do the same. This reminds us of the segmentation of cell life observed under the microscope, as though a four-dimensional figure were registering its passage through our space.

THE ELECTRIC CURRENT

Hinton conceived of an electric current as a four-dimensional vortex. He declared that on the Higher Space Hypothesis the revolution of the ether would yield the phenomenon of the electric current. The reader is referred to Hinton's book, *The Fourth Dimension,* for an extended development of this idea. What follows is a brief summary of his argument. First, he examines the characteristics of a vortex in a three-dimensional fluid. Then he conceives of what such a vortex would be in a four-dimensional medium of analogous properties. The whirl would be about *a plane,* and the contour of this plane would correspond to the ends of the axis line in the former vortex; and as therein, the vortex would extend to the boundary of the medium. Every electric current forms a closed circuit: this is equivalent to the hyper-vortex having its ends in the boundary of the hyper-fluid. The vortex with *a surface* as its axis, therefore, affords a geometric image of a closed circuit.

Hinton supposes a conductor to be a body which has the property of serving as a terminal abutment to such a hyper-vortex as has been described. The conception that he forms of a closed current, therefore, is of a vortex sheet having its *edge* along the circuit of the conducting wire. The whole wire would then be like the centres on which a spindle turns in three-dimensional space, and any interruption of the continuity of the wire would produce a *tension* in place of a continuous revolution. The phenomena of electricity—polarity, induction, and the like—are of the nature of the stress and strain of a medium, but one possessing properties unlike those of ordinary matter. The phenomena lend themselves to explanation in terms of higher space. If Hinton's hypothesis be the true explanation, then the universality of electromagnetic action would again point to the conclusion that our three-dimensional world is *superficial*—the surface, that is, of a four-dimensional universe.

THE GREATER UNIVERSE

This practically exhausts the list of accepted and accredited indications of hyper-dimensionality in our physical environment. But if the collective human consciousness is moving into the fourth dimension such indications are bound to multiply out of all measure. It should be remembered that in Franklin's day electricity was manifest only in the friction of surfaces and in the thunderbolt. Today all physical phenomena, in their last analysis, are considered to be electrical. The world is not different but perception has evolved—and is evolving.

There is another field in which some of our ablest minds are searching for evidences of the curvature of space: the field of astronomy and astro-physics. But into this the layman hesitates to enter because the experts themselves have found no common ground of understanding. The ether of space is a battlefield strewn with dead and dying hypotheses; gravitation, like multiplication, is vexation; the very nature of time, form and movement is under vivid discussion, in connection with the Theory of Relativity.

Notwithstanding these counter-currents of speculation, which should make the wise man speak smilingly of his wisdom, this summary remains incomplete without a reference to the pressure of higher space upon those adventurous minds that essay to deal with the profound problems of the greater universe, and a statement of the reasons for their feeling this pressure. These reasons are well suggested by Professor B. G. Harrison, in his *Popular Astronomy*. He says: "With the idea of a universe of finite dimensions there is the obvious difficulty of the beyond. The truth is that a universe of finite proportions is equally difficult to realize as one of infinite extent. Perhaps the nearest analogy to infinity that we can understand lies in our conception of a closed curve. It seems easier to imagine the endless movement of a sphere in a circular path than the case of one traveling in a straight line. Possibly this analogy may apply in some way to fourth-dimensional space, but the manner of its application is certainly not easy to understand. If we would imagine that all co-ordinates of time and space were curved, and eventually return to the same point, it might bring the ultimate comprehension one degree nearer."

A HINT FROM ASTRONOMY

The physical evidence that our space is thus curved in higher space, some have considered astronomy to furnish in what is called the "negative parallax" of certain distant stars. This cannot be passed by, though it is too deeply involved with the probable error of the observers themselves to be considered more than an interesting fact in this connection.

Everyone knows that the difference of angle under which an object is seen from two standpoints is called its parallax. The parallax of the stars—and the consequent knowledge of their distance—is obtained by observing them from opposite points of the earth's orbit around the sun. When a star is within measurable distance, these angles are acute, and the lines from the star to the earth at opposite sides of its orbit converge, therefore. But when these lines, as sometimes happens, appear to be *divergent,* the result is called a *negative* parallax, and is explainable by higher space relationships. Obviously, the divergence of the lines would indicate that the object lies *behind* the observer instead of in front of him. This anomaly can be explained by the curvature of space in the fourth dimension. If space is so curved, the path of light itself is curved also, and a man—were his vision immeasurably keen, not to say telescopic—could see the back of his own head! It is not worth while to give this question of negative parallax too much importance, by reason of the probability of error, but in this connection it should be stated that there appears to be an undue number of negative parallaxes recorded.

GRAVITATION

Gravitation remains a puzzle to science. The tendency of modern physics is to explain all material phenomena in terms of electrons and the ether, but the attempt to account for gravitation in this way is attended with difficulties. In order to cope with these, it seems necessary to assume that our universe is only a portion of a greater universe. This assumption readily lends itself to the conception of our universe as a three-dimensional meeting place of two portions of a universe of four dimensions—that is, its conception as a "higher" surface. This is a fundamental postulate of higher space speculation.

One hypothesis advanced to explain gravitation assumes the existence of a constant hydrostatic pressure transmitted through the ether. A steady flow of ether into every electron in a gravitating system of bodies would give rise to forces of attraction between them, varying inversely as the square of the distance, according to Newton's law. But in order to avoid the conception of the continual destruction and creation of ether, it is necessary to assume a steady flow through every electron between our universe and the greater universe of which it is assumed to form a part. Now because the electrons, in order to receive this flow, must lie on the boundary of this greater universe, the latter must be four-dimensional. Every electron, in other words, must be the starting point of a pathway into, and a terminal point out of, four-dimensional space. Here we have another familiar higher space concept.

THE ETHER OF SPACE

The ether of space, because it has at last found entrance, must be given a grudging hospitality in these pages, even though the mysterious stranger prove but a ghost. The Relativists would have it that with the acceptance of their point of view the ether may be eliminated; but if they take away the ether they must give us something in its stead: in whatever way the science of the future disposes of this problem, it must take into account the fact of light transmission. On the theory that the ether is an elastic solid of amazing properties, in which the light waves vibrate transversely to their direction, it assists the mind to think of the ether as four-dimensional, because then a light-wave would be a superficial disturbance of the medium—superficial, but three-dimensional, as must needs be the case with the surface of a four-dimensional solid.

This search for evidences of hyper-dimensionality in the universe accessible to our senses is like looking, not for a needle in a haystack, but for a haystack in a needle—for the greater in the less. From the purely physical evidences, all that can with certainty be said is that the hypothesis is not inconsistent with the facts of science or its laws; that it is being verified and rendered more probable by the investigations of science; that it is applicable to the description or explanation of all the observed phenomena, and assigns a cause fully adequate to have produced them.

Now there is an order of phenomena that we call psychic. Because they are phenomenal they cannot occur outside of time and space altogether; because they are psychic they defy explanation in terms of the space and time of every-day life. Let us next examine these in the light of our hypothesis.

IV

TRANSCENDENTAL PHYSICS

ZÖLLNER

IN the year 1877, Johann Friedrich Zöllner, professor of physics and astronomy at the University of Leipsic, undertook to prove that certain (so-called) psychic phenomena were susceptible of explanation on the hypothesis of a four-dimensional space. He used as illustrations the phenomena induced by the medium Henry Slade. By the irony of events, Slade was afterward arrested and imprisoned for fraud, in England. This fact so prejudiced the public mind again Zöllner that his name became a word of scorn, and the fourth dimension a synonym for what is fatuous and false. Zöllner died of it, but since his death public opinion has undergone a change. There is a great and growing interest in everything pertaining to the fourth dimension, and belief in that order of phenomena upon which Zöllner based his deductions is supported by evidence at once voluminous and impressive.

It is unnecessary to go into the question of the genuineness of the particular phenomena which Zöllner witnessed. His conclusions are alone important, because they apply equally to other manifestations, whose authenticity has never been successfully impeached. Zöllner's reasoning with regard to certain psychic phenomena is somewhat along the following lines.

APPARITIONS

The intrusion (as an apparition) of a person or thing into a completely enclosed portion of three-dimensional space; or contrariwise, the exit (as an evanishment) out of it.

Because we lack the sense of four-dimensional space, we must here have recourse to analogy, and assume three-dimensional space to be the unsensed higher region encompassing a world of two dimensions. To a hypothetical flat-man of a two-space, any portion of his plane surrounded by an unbroken line would constitute an enclosure. Were he confined within it, escape would be impossible by any means known to him. Had he the ability to move in the third dimension, however, he could rise, pass over the enclosing line without disturbing it, and descend on the other side. The moment he forsook the plane he would disappear from two-dimensional space. Such a disappearance would constitute an occult phenomenon in a world of two dimensions.

Correspondingly, an evanishment from any three-dimensional enclosure—such as a room with locked doors and windows—might be effected by means of a movement in the fourth dimension. Because a body would disappear from our perception the moment it forsook our space, such a disappearance would be a mystery; it would constitute an occult phenomenon. The thing would be no more mysterious, however, to a consciousness embracing four dimensions within its ken, than the transfer of an object from the inside to the outside of a plane figure without crossing its linear boundary is mysterious to us.

POSSESSION

The temporary possession of a person's body, or some member of that body, by an alien will, as exemplified in automatic writing and obsession.

It would doubtless amaze the scientifically orthodox to know how many people habitually and successfully practise the dubious art of automatic writing—not mediums, so-called, but people of refinement and intelligence. Although the messages received in this way may emanate from the subconscious mind of the performer, there is evidence to indicate that they come sometimes from an intelligence discarnate, or from a person remote from the recipient in space.

If such is indeed the case, if the will is extraneous, how does it possess itself of the nerves and muscles of the hand of the writer? The Higher Space Hypothesis is of assistance here. It is only necessary to remember that from the fourth dimension the interior of a solid is as much exposed as the interior of a plane figure is exposed from the region of the third dimension. A four-dimensional being would experience no difficulty, under suitable conditions, in possessing itself of any part of the bodily mechanism of another.

The same would hold true in cases of possession and obsession; for if the bastion of the hand can thus be captured, so also may the citadel of the brain. Certain familiar forms of hypnotism are not different from obsession, the hypnotizer using the brain and body of his subject as though they were his own. All unconsciously to himself, he has called into play four-dimensional mechanics. Many cases of so-called dual personality are more easily explicable as possession by an alien will, than on the less credible hypothesis that the character, habits, and language of a person can change utterly in a moment of time.

CLAIRVOYANCE IN SPACE

Vision at a distance and the exercise of a superior power of sight.

Clairvoyance in space is of various kinds and degrees. Sometimes it consists in the perception of super-physical phenomena—the upfurling of a curtain on a strange and wonderful world; and again it appears to be a higher power of ordinary vision, a kind of seeing to which the opacity of solids offers no impediment, or one involving spatial distances too great and too impeded for normal physical vision to be effective.

That clairvoyance which consists in the ability to perceive not alone the superficies of things as ordinary vision perceives them, but their interiors as well, is analogous to the power given by the X-ray, by means of which, on a fluorescent screen, a man may behold the beating of his own heart. But if the reports of trained clairvoyants are to be believed there is this difference: everything appears to them without the distortions due to perspective, objects being seen as though they were inside and not outside of the perceiving organ, or as though the observer were in the object perceived, or in all places at the same time.

Our analogy makes all this intelligible. To the flat-man, clairvoyance in space would consist in that power of perception which we exercise in reference to his plane. From the third dimension the boundaries of plane figures offer no impediment to the view of their interiors, and they themselves in no way impede the sight of surrounding objects. If we assume that clairvoyance in space is the perception of the things of our world from the region of the fourth dimension, the phenomena exactly conform to the demands of our analogy. It is no more difficult for a four-dimensional intelligence to understand the appearance or disappearance of a body in a completely closed room, or the withdrawal of an orange from its skin, without cutting or breaking that skin, than it is for us to see the possibility of taking up a pencil point from the centre of a circle and putting it down outside. We are under no compulsion to draw a line across the circumference of the circle for the point to enter or leave it. Moreover the volume of our sensible universe embraced in the clairvoyant's field of vision will increase in the same way that a balloonist's view increases in area as he rises above the surface of the earth. To account for clairvoyant vision at a distance, it is of course necessary to posit some perceptive organ other than the eye, but the fact that in trance the eyes are closed, itself demands this assumption.

CLAIRVOYANCE IN TIME

The perception of a past event as in process of occurring, or the pre-vision of something which comes to pass later.

No mechanistic explanation will serve to account for this order of clairvoyance, since it is inextricably involved in the mystery of consciousness itself. Yet our already overworked analogy can perhaps cast a little light even here.

To the flat-man the third dimension of objects passing through his plane translates itself to his experience into *time.* Were he capable of rising in the positive direction of the third dimension, he would have pre-vision, because he would be cognizant of that which had not yet intersected his plane; by sinking in the negative direction, he would have post-vision, because he could recognize that which had already passed.

Now there are excellent reasons, other than those based on analogy, why the fourth-dimensional aspect of things may manifest itself to our ordinary experience, not as spatial extension, but as temporal change. Then, if we conceive of clairvoyance as a transcending by consciousness of our three-dimensional space, pre-vision and post-vision would be logically possible as corresponding to the positive and negative of the fourth dimension. This may be made clearer by the aid of a homely illustration.

PISGAH SIGHTS OF LIFE'S PAGEANT

Suppose you are standing on a street corner watching a procession pass. You see the pageant as a sequence of objects and individuals appearing into view near by and suddenly, and disappearing in the same manner. This would represent our ordinary waking consciousness of what goes on in the world round about. Now imagine that you walk up the street in a direction opposite to that in which the procession is moving. You then rapidly pass in review a portion of the procession which had not yet arrived at the point where you were immediately before. This would correspond to the seeing of something before it "happened," and would represent the positive aspect of clairvoyance in time—pre-vision. Were you to start from your original position, and moving in the direction in which the procession was passing, overtake it at some lower street corner, you could witness the thing you had already seen. This would represent post-vision—clairvoyance of the past.

A higher type of clairvoyance would be represented by the sweep of vision possible from a balloon. From that place of vantage the procession would be seen, not as a sequence, but simultaneously, and could be traced from its formation to its dispersal. Past, present and future would be merged into one.

It is true that this explanation raises more questions than it answers: to account in this way for a marvel, a greater marvel must be imagined—that of transport out of one's own "space." The whole subject bristles with difficulties, not the least of which is that even to conceive of such a thing as pre-vision all our old ideas about time must be recast. This is being done in the Theory of Relativity, a subject which may appropriately engage our attention next.

V

CURVED TIME

TIME FROM THE STANDPOINT OF EXPERIMENT AND OF CONSCIOUS EXPERIENCE

IN some moment of "sudden light" what one of us has not been able to say, with Rossetti,

> "I have been here before,
> But when or how I cannot tell."

Are such strange hauntings of our House of Life due to the cyclic return of time? Perhaps—but what is time?

Suppose some one should ask you, "What is an hour?" Your answer might be, "It is the interval marked off by the clock-hand between 1 and 2." "But what if your clock is running down or speeding up?" To this you would probably reply, "The clock is set and corrected by the earth, the sun and the stars, which are constant in their movements." *But they are not.* The earth is known to be running slow, by reason of tide friction, and this is likely to continue until it will revolve on its axis, not once a day, but once a year, presenting always the same face to the sun.

We can only measure time by *uniform* motion.

Observe the vicious circle. Uniform motion means the covering of equal spaces in equal times. But how are we to determine our equal times? Ultimately we have no other criterion save the uniform motion of the clock-hand or the star-dial. The very expressions, "uniform motion," "equal times," beg the whole question of the nature of time.

Let us then, in this predicament, consider time not from the standpoint of experiment, but of conscious experience—what Bergson calls "real duration."

Every point along the line of memory, of conscious experience, has been traced out by that unresting stylus we call "the present moment." The question of its rate of motion we shall not raise, as it is one with which we have found ourselves impotent to deal. We believe on the best of evidence that the conscious experience of others is conditioned like our own. For better understanding let us have recourse to a homely analogy: let us think of these more or less parallel lines of individual experience in the semblance of the strands of a skein of flax. Now if, *at the present moment,* this skein were cut with a straight knife at right angles to its length, the cut end would represent *the time plane*—that is, the present moment of all—and it would be the same for all, provided that the time plane were flat. *But is it really flat?* Isn't the straightness of the knife a mere poverty of human imagination? Existence is always richer and more dramatic than any diagram.

> "Line in nature is not found;
> Unit and universe are round.
> In vain produced, all rays return;
> Evil will bless and ice will burn."

Undoubtedly the flat time-plane represents with fair accuracy the temporal conditions that obtain in the human aggregate in this world under normal conditions of consciousness, but if we consider our relation to supernormal states, or to intelligent beings upon distant worlds of the visible universe the conditions might be widely different. The time-section corresponding to what our straight knife made flat in the case of the flax may be—nay, probably is—strongly curved.

RELATIVITY

This crude analogy haltingly conveys what is meant by curved time. It is an idea which is implicit in the Theory of Relativity. This theory has profoundly modified many of our basic conceptions about the universe in which we are immersed. It is outside the province of this book and beyond the power of its author even so much as to sketch the main outlines of this theory, but certain of its conclusions are indispensable, since they baldly set forth our dilemma in regard to the measurement of space and time. We can measure neither, except relatively, because they must be measured one by the other, and no matter how they vary, these variations always compensate one another, leaving us in the same state of ignorance that we were in before.

Suppose that two intelligent beings, one on Mars, let us say, and the other on the earth, should attempt to establish *the same moment of time* by the interchange of light signals, or by any other method which the most rigorous science could devise. Assume that they have for this purpose two identically similar and mechanically perfect chronometers, and that every difficulty of manipulation were successfully overcome. Their experiment could end only in failure, and the measure of this failure neither one, in his own place, could possibly know. If, after the experiment, the Martian, chronometer in hand, could be instantly and miraculously transported to the earth, and the two settings compared, they would be found to be different: how different, we do not know.

The reason for the failure of any such experiment anywhere conducted can best be made plain by a crude paraphrase of a classic proposition from Relativity. Suppose it is required to determine the same moment of time at two different places on the earth's surface, as must be attempted in finding their difference in longitude. Take the Observatory at Greenwich for one place, and the Observatory at Washington for the other. At the moment the sun is on the meridian of Greenwich, the exact time of crossing is noted and cabled to Washington. The chronometer at Washington is set accordingly, and the time checked back to Greenwich. This message arrives two seconds, say, after the original message was sent. Washington is at once notified of this double transmission interval. On the assumption that HALF of it represents the time the message took to travel from east to west, and the other half the time from west to east again, the Washington chronometer is set one second ahead of the signalled time, to compensate for its part of the loss. When the sun has reached the meridian of Washington, the whole process is repeated, and again, as before, half of the time the message has taken to cross and recross the Atlantic is added to the Greenwich record of noon at Washington. The number of hours, minutes, seconds and fractions of a second between these two corrected records represents the difference in solar time between the two places, and incidentally the same moment of time has been established for both—at least so it would appear.

But is it established? That each message took an equal time to travel each way is a pure assumption, and happens to be a false one. The accuracy of the result is vitiated by a condition of things to which the Relativists have called attention. Our determination might be defended if Washington and Greenwich could be assumed to remain at rest during the experiments, and some argument might even be made in its favour if we could secure any cosmic assurance that the resultant motion of the earth should be the same when Greenwich signalled its noon to Washington and Washington its noon to Greenwich.

Our present discussion is merely illustrative, or diagrammatic; so we will neglect the velocity of the earth in its orbit around the sun, some forty times greater than that of a cannon ball, and the more uncertain and more vertiginous speed of the whole solar system toward its unknown goal. Let us consider only the rotation of the earth on its axis, the tide-speed of day and night. To fix our idea, this may be taken, in our latitudes, at eighteen thousand miles per day, or perhaps half the speed of a Mauser rifle bullet.

So fast, then, will Washington have been moving to meet the message from Greenwich. So fast will Greenwich have been retreating from Washington's message.

Now the ultimate effect of motion on the time-determination cannot be calculated along any such simple lines as these. Indeed, it cannot be exactly calculated at all, for we have not all the data, but there is certainly *some* effect. Let us take another example. Suppose one rows four miles up a river against a current of two miles per hour, at a rowing speed of four miles per hour. This will take two hours, plainly. The return trip with the river's gift of two miles per hour will evidently require but forty minutes. *Two hours and forty minutes* for the round trip of eight miles, therefore. Now to row eight miles in still water, according to our supposition, would have required but *two hours.* But, some one objects, the current must help the return trip as much as it hindered the outgoing! Ah, here is the snare that catches rough-and-ready common sense! How long would the double journey have taken *if the river current had been faster than our rowing speed?* How shall we schedule our trip if we cannot learn the correct speed, *or if it varies from minute to minute?*

The inferences which we may draw from our hypothetical experiment are plain. The settings of the two chronometers would be defective: they would not show the same time, but each of them would mark the *local* time, proper to its own place. There would be no means of detecting the amount of error, since the messages were transmitted by a medium involved with them in their transportation. If only local time can be established, the possibility of a warped time-plane—the curvature of time—is directly opened up.

Doubtless it is true that on so relatively minute a scale as is offered by the earth, any deviation from perfect flatness of the time-plane would be so inconsiderable and imperceptible as to make it scientifically negligible; but this by no means follows when we consider our relation to other worlds and other systems.

A similar condition holds with regard to space-distortion. The Theory of Relativity enforces the conclusion that from the standpoint of our conventions in regard to these matters, all bodies involved in transportation undergo a contraction in the direction of that transportation, while their dimensions perpendicular to the transportation remain invariable. This contraction is the same for all bodies. For bodies of low velocity, like the earth, this distortion would be almost immeasurably slight; but great or little, no measuring instruments on the body transporting would ever disclose it, for a measure would undergo the same contraction as the thing measured.

The above explanations are necessarily symbolistic rather than demonstrative, but any one who will seriously follow out these lines of thought, or, still better, study the attitude of the hard-headed modern physicist towards our classical geometry and mechanics, cannot fail to realize how conventional, artificial—even phantasmal—are the limitations set by the primitive idea of flat space and straight time.

THE SPOON-MAN

These concepts, that space and time are not as immutable as they appear, that our universe may suffer distortion, that time may lag or hasten without our being in the least aware, may be made interestingly clear by an illustration first suggested by Helmholtz, of which the following is in the nature of a paraphrase.

If you look at your own image in the shining surface of a teapot, or the back of a silver spoon, all things therein appear grotesquely distorted, and all distances strangely altered. But if you choose to make the bizarre supposition that this spoon-world is real, and your image—the spoon-man—a thinking and speaking being, certain interesting facts could be developed by a discussion between yourself and him.

You say, "Your world is a distorted transcript of the one in which I live."

"Prove it to me," says the spoon-man.

With a foot-rule you proceed to make measurements to show the rectangularity of the room in which you are standing. Simultaneously he makes measurements giving the same numerical results; for his foot-rule shrinks and curves in the exact proportion to give the true number of feet when he measures his shrunken and distorted rear wall. No measurement you can apply will prove you in the right, or him in the wrong. Indeed he is likely to retort upon you that it is your room which is distorted, for he can show that in spite of all its nightmare aspects his world is governed by the same orderly geometry that governs yours.

The above illustration deals purely with space relations, for such relations are easily grasped, but certain distortions in time relations are no less absolutely imperceptible and unprovable. So far from having any advantage over the spoon-man, our plight is his. The Principle of Relativity discovers us in the predicament of the Mikado's "prisoner pent," condemned to play with crooked cues and elliptical billiard balls; and of the opium victim for whom "space swells," and time moves sometimes swift and sometimes slow.

THE ORBITAL MOVEMENT OF TIME

Now if our space is curved in higher space, since such curvature is at present undetectable by us, we must assume, as Hinton chose to assume, that it curves in the minute, or, as some astronomers assume, that its curve is vast. These assumptions are not mutually exclusive: they are quite in analogy with the general curvature of the earth's surface which is in no wise interfered with by the lesser curvatures represented by mountains and valleys. It is easiest to think of our space as completely curved in higher space in analogy with the surface of a sphere.

Similarly, if time is curved, the idea of the cyclic return of time naturally (though not inevitably) follows, and the division of the greater cycles into lesser loops; for it is easier to assign this elliptical movement to time than any other, by reason of the orbital movements of the planets and their satellites. What results from conceptions of this order? Amazing things! If our space is curved in higher space, you may be looking toward the back of your own head. If time flows in cycles, in travelling toward tomorrow you may be facing yesterday.

This "eternal return," so far from being a new idea, is so old that it has been forgotten. Its reappearance in novel guise, along with so many other recrudescences, itself beautifully illustrates time curvature in consciousness. *Yugas,* time cycles, are an integral and inexpugnable part of Oriental metaphysics. "Since the soul perpetually runs," says Zoroaster, "in a certain space of time it passes through all things, which circulation being accomplished, it is compelled to run back again through all things, and unfold the same web of generation in the world." Time curvature is implicit in the Greek idea of the iron, bronze, silver, and golden ages, succeeding one another in the same order: the winter, seed-time, summer and harvest of the larger year. Astrology, seership, prophecy, become possible on the higher-time hypothesis. From this point of view history becomes less puzzling and paradoxical. What were the Middle Ages but a forgetting of Greek and Roman civilization, and what was the Renaissance but a remembering of them—a striving to re-create the ruined stage-settings, and to re-enact the urbane play of Pagan life?

But it is in the individual consciousness that time curvature receives its most striking confirmation—in those lesser returns and rhythms to which we give the name of *periodicity*. Before considering these, however, a fundamental fallacy of the modern mind must be exposed.

MATERIALITY THE MIRROR OF CONSCIOUSNESS

Our vicious habit of seeking the explanation of everything—even thought and emotion—in materiality, has betrayed us into the error of attributing to organic and environic changes the very power by which they are produced. We are wont to think of feeling, the form in which Being manifests to consciousness, as an effect instead of as a cause. When Sweet Sixteen becomes suddenly and mysteriously interesting to the growing boy, it is not because sex has awakened in his body, but because the dread time has come for him to contemplate the Idea of Woman in his soul. If you are sleepy, it is not because the blood has begun to flow away from your brain, but because your body has begun to bore you. Night has brought back the Idea of Freedom, and consciousness chloroforms the thing that clutches it. If you are ill, you grow cold or your temperature rises: it is the signal by which you know that your consciousness is turning toward the Idea of Pain.

Just as a savage looks for a man behind the mirror, we foolishly seek in materiality for that which is not there. The soul determines circumstance: the soul contains the event which shall befall. The organic and environic rearrangements incident to obscure rotations in higher space are like the changes a mirror-image undergoes as an object draws near and then recedes from its plane. This is only a figure of speech, but it is susceptible of almost literal application. Ideas, emerging from the subconscious, approach, intersect, recede from, and re-approach the stream of conscious experience; taking the forms of aversions and desires, they register themselves in action, and by reason of time curvature, everything that occurs, recurs.

PERIODICITY

We recognize and accept this cyclic return of time in such familiar manifestations of it as Nature affords in *periodicity*. We recognize it also in our mental and emotional life, when the periods can be co-ördinated with known physical phenomena, as in the case of the wanderlust which comes in the spring, the mild melancholy of autumn, the moods that go with waning day, and winter night. It is only when these recurrences do not submit themselves to our puny powers of analysis and measurement that we are incredulous of a larger aspect of the law of time-return. Sleep, for example, is not less mysterious than death, which too may be but "a sleep and a forgetting." The reason that sleep fails to terrify us as death does is because experience has taught that *memory leaps the chasm.* Why should death be dreaded any more than bedtime? Because we fear that we shall forget. But do we really ever forget? As Pierre Janet so tersely puts it: "Whatever has gone into the mind may come out of the mind"—and in a subsequent chapter this aphorism will be shown to have extension in a direction of which the author of it appears not to have been aware. Memory links night to night and winter to winter, but such things as "the night-time of the spirit" and "the winter of our discontent" are not recognized as having either cause or consequence. Now though the well-springs of these states of consciousness remain obscure, there is nothing unreasonable in believing that they are recrudescences of far-off, forgotten moods and moments; neither is it absurd to suppose that they may be related to the movements and positions of the planets, as night and winter are related to the axial and orbital movements of the earth.

But there are other and even more interesting evidences of time curvature in consciousness. These lead away into new regions which it is our pleasure now to explore.

VI

SLEEP AND DREAMS

SLEEP

OUR space is called three-dimensional because it takes three numbers—measurement in three mutually perpendicular directions—to determine and mark out any particular point from the totality of points. Time, as the individual experiences it, is called one-dimensional for an analogous reason: one number is all that is required to determine and mark out any particular event of a series from all the rest. Now in order to establish a position in a space of four dimensions it would be necessary to measure in *four* mutually perpendicular directions. Time curvature opens up the possibility of a corresponding higher development in time: one whereby time would be more fittingly symbolized by a plane than by a linear figure. Indeed, the familiar mystery of memory calls for such a conception. Memory is a carrying forward of the past into the present, and the fact that we can recall a past event without mentally rehearsing all the intermediate happenings in inverse order, shows that in the time aspect of memory there is simultaneity as well as sequence—time, if we choose to figure it as one-dimensional, ceases to be *linear* and becomes *plane.* More remarkable illustrations of the sublimation of the time-sense are to be found in the phenomena of sleep and dreams.

"Oh, thou that sleepest, what is sleep?" asks the curious Leonardo. Modern psychological science has little to offer of a positive nature in answer to this world-old question, but it has at least effectively disposed of the absurd theories of the materialists who would have us believe that sleep is a mere matter of blood circulation or of intoxication by an accumulation of waste products in the system. Sleep states are not abnormal, but part and parcel of the life experience of the individual. When a person is asleep he has only become unresponsive to the mass of stimuli of the external world which constitutes his environment. As Sidis says, "When our interest in external existence flags and fades away, we go to sleep. When our interests in the external world cease, we draw up the bridges, so to say, interrupt all external communication as far as possible, and become isolated in our own fortress and repair to our own world of organic activity and inner dream life. Sleep is the interruption of our intercourse with the external world: it is the laying down of our arms in the struggle of life. Sleep is a truce with the world."

The twin concepts of higher space and curved time sanction a view of sleep even bolder. Sleep is more than a longing of the body to be free of the flame which consumes it: the flame itself aspires to be free —that is to say, consciousness, tiring of its tool, the brain, and of the world, its workshop, takes a turn into the plaisance of the fourth dimension, where time and space are less rigid to resist the fulfilment of desire.

DREAMS

We find a confirmation of this view in dream phenomena. But however good the evidence, we shall fail to make out a case unless dream experiences are conceded to be as real as any other. Any reluctance we may have, to make this concession, comes in the first place from the purely subjective character of dreams, and in the second place from their triviality and irrationality—it is as though the muddy sediment of daytime thought and feeling, and that alone, were there cast forth.

In answer to the first objection, advanced psychology affirms that the subconscious mind, from which dreams arise, approaches more nearly to the omniscience of true being than the rational mind of waking experience. The triviality and irrationality of dreams are sufficiently accounted for if the dream state is thought of as the meeting place of two conditions of consciousness: the foam and flotsam "of perilous seas in faëry lands forlorn," whose vastitude, whose hidden life, and rich argosies of experience can be inferred only from the fret of the tide on their nether shore—the tired brain in sleep.

For it is the *remembered* dream alone that is incoherent—the dream that comes clothed in the rags and trappings of this work-a-day world, and so leaves some recoverable record on the brain. We all feel that the dreams we cannot remember are the most wonderful. Who has not awakened with the sense of some incommunicable experience of terror or felicity, too strange and poignant to submit itself to concrete symbolization, and so is groped for by the memory in vain? We know that dreams grow more ordered and significant as they recede from the surface of consciousness to its depths. Deep sleep dreams are in the true sense clairvoyant, though for the most part irrecoverable—"Canst thou draw out Leviathan with an hook?" DuPrel and others have shown that the difference between ordinary dreaming, somnambulance, trance and ecstasy, is only a matter of redistribution of thresholds—that they are all related states and merge into one another. We have therefore every right to believe that for a certain number of hours out of the twenty-four we are all sibyls and seers, however little most of us are able to profit by it. Infrequently, in moments of peculiar susceptibility, the veil is lifted, but the art of *dreaming true* remains for the most part unmastered—one of the precious gifts which the future holds in store for the sons and daughters of men.

The partial waking state is the soil in which remembered dreams develop most luxuriously. Paradoxical as it may sound, they are the product not of our sleep but of our waking. Such dreams belong to both worlds, partly to the three-dimensional and partly to the four-dimensional. While dreams are often only a hodge-podge of daytime experience, their incredible rapidity, alien to that experience, gives us our first practicable intimation of a higher development of time.

TIME IN DREAMS

The unthinkable velocity of time in dreams may be inferred from the fact that between the moment of impact of an impression at the sense-periphery and its reception at the centre of consciousness—moments so closely compacted that we think of them as simultaneous—a coherent series of representations may take place, involving what seem to be protracted periods for their unfoldment. Every reader will easily call to mind dream experiences of this character, in which the long-delayed dénouement was suggested and prepared for by some extraneous sense-impression, showing that the entire dream drama unfolded within the time it took that impression to travel from the skin to the brain.

Hasheesh dreams, because they so often occur during some momentary lapse from normal consciousness and are therefore measurable by its time scale, are particularly productive of evidence of the "looping" of time. Fitzhugh Ludlow, in *The Hasheesh Eater,* narrates the dreams that visited him in the brief interval between two of twenty or more awakenings on his walk homeward after his first experience with the drug. He says, "I existed by turns in different places and various states of being. Now I swept my gondola through the moonlit lagoons of Venice. Now Alp on Alp towered above my view, and the glory of the coming sun flashed purple light upon the topmost icy pinnacle. Now in the primeval silence of some unexplored tropical forest I spread my feathery leaves, a giant fern, and swayed and nodded in the spice-gales over a river whose waves at once sent up clouds of music and perfume. My soul changes to a vegetable essence, thrilled with a strange and unimagined ecstasy."

Earlier in the same evening, when he was forced to keep awake in order not to betray his condition, the dream time-scale appears to have imposed itself upon his waking consciousness with the following curious effect. A lady asked him some question connected with a previous conversation. He says,

"As mechanically as an automaton I began to reply. As I heard once more the alien and unreal tones of my own voice, I became convinced that it was some one else who spoke, and in another world. I sat and listened: still the voice kept speaking. Now for the first time I experienced that vast change which hasheesh makes in all measurements of time. The first word of the reply occupied a period sufficient for the action of a drama; the last left me in complete ignorance of any point far enough back in the past to date the commencement of the sentence. Its enunciation might have occupied years. I was not in the same life which had held me when I heard it begun."

This well-known fact, that we cannot measure dreams by our time scale, proves that time in sleep does not correspond with in waking, and that the "dream organ" of consciousness has a time scale of its own. If in our waking state we experience one kind of time, and in dreams quite another, the solution of the mystery should be sought in the *vehicle* of consciousness, for clearly the limit of impressionability or power of response of the vehicle establishes the time-scale, just as the size of the body with relation to objects establishes the space-scale. Time must be different for the ant and the elephant, for example, as space is different.

Our sense of time is wholly dependent upon the

rapidity with which impressions succeed one another. Were we capable of receiving only one impression an hour, like a bell struck every hour with a hammer, the ordinary term of life would seem very short. On the other hand, if our time-sense were always as acute as it is in dreams, uncounted æons would seem to be lived through in the interval between childhood and old age.

Imagine a music machine so cunningly constructed and adjusted as not only to sound each note and chord in its proper sequence and relation, but to regulate also the duration of the sound vibration. If this machine were operated in such a manner as to play, in a single second of time, the entire overture of an opera which would normally occupy half an hour, we should hear either nothing at all or only an unintelligible noise a second long. This would be due to no defect in the *sound-producing* mechanism, but to the limitations of the *sound-receiving* mechanism, our auditory apparatus. Could this be altered to conform to the unusual conditions—could it capture and convey to consciousness every note of the overture in a second of time—that second would seem to last half an hour, provided that every other criterion for the measurement of duration were denied for the time being.

Now dreams *seem* long: we only discover afterwards and by accident their almost incredible brevity. May we not—must we not—infer from this that the body is an organ of many stops and more than one keyboard, and that in sleep it gives forth this richer music, that the House of Sleep is higher-dimensional both in its space and its time aspects?

This theory of a higher-dimensional existence during sleep accounts in part for the great longing for sleep. "What is it that is much desired by men, but which they know not while possessing?" again asks Leonardo. *"It is sleep,"* is his answer. This longing for sleep is more than a physical longing, and the refreshment it brings is less of the flesh than of the spirit. It is possible to withstand the deprivation of food and water longer and better than the deprivation of sleep, and its recuperative power is correspondingly greater.

Experiments have been made with mature University students by which they have been kept awake ninety-six hours. When the experiments were finished the young men were allowed to sleep themselves out, until they felt they were thoroughly rested. All awoke from a long sleep completely refreshed, but the one who took longest to restore himself from his protracted vigil slept only one-third longer than was regular with him. And this has been the experience over and over again of men in active life who have been obliged to keep awake for long periods by the absolute necessities of the situation in which they have been placed.

In this fact there is surely another hint of the sub-limitation of the time-sense during sleep. While it would be an unwarrantable assumption to suppose that the period of recuperation by sleep must be as long, or nearly as long, as the period of deprivation, the ratio between the two presents a discrepancy so great that it would seem as though this might be due to an acceleration of the time element of consciousness.

THE EASTERN TEACHING IN REGARD TO SLEEP AND DREAMS

In this matter of the wonder, the mystery, the enchantment of sleep and dreams, the most modern psychology and the most ancient wisdom meet on common ground. Eastern wisdom casts such a light upon the problems of subjectivity that it should not be lightly dismissed. For uncounted centuries Hindu-Aryan spiritual science has recognized, not one plane or condition of consciousness, but three: waking, dreaming, and deep sleep—the gross, the subtle and the pure. In the waking state—that is, with the vehicle attuned to vibrate to materiality—the individual self is as a captive in a citadel of flesh, aware of only so much of the universal life as chances to enact itself before the windows of his prison. In the dream state, when the more violent vibrations of the body are stilled in sleep, consciousness becomes active in its subtle (four-dimensional) vehicle, and ranges free throughout the ampler spaces of this subtler world. In deep sleep consciousness reverts to its pure condition—the individual self becomes the All-Self: the rainbow, no longer prismatic by reason of its refraction in materiality, becomes the pure white light; the melody of life resolves itself into the primordial harmony; sequence becomes simultaneity, and Time, no longer "besprent with seven-hued circumstance," is swallowed up in duration.

"There are two paths for him, within and without, and they both turn back in a day and a night. . . . After having subdued by sleep all that belongs to the body, he, not asleep himself, looks down upon the sleeping. Having assumed light, he goes again to his place, the golden person, the lonely bird."

SPACE IN DREAMS

However preposterous may appear to us this notion that the waking state, in which we feel ourselves most potent and alive, is really one of inhibition—that the world is only a "shoal of time"—it is curiously borne out by the baffling phenomena of dreams, and is in perfect accord with the Higher Space Hypothesis. The possibility of shaking off the grip of sleep under appropriate circumstances, the fact that we can watch in our sleep, and awake at the right moment, that we can sleep and still watch and keep awake in regard to special objects and particular persons—these things raise insuperable difficulties against all those plausible and apparently scientific theories of sleep current in the West; but they fit perfectly with the Eastern idea that "he, not asleep himself, looks down upon the sleeping." And to the question, "How, and from whence?" in the light of our hypothesis we may answer, "By the curvature of time, consciousness escapes into the fourth dimension."

Myers shows that he was in need of just this clue in order to account for some of the dream experiences recorded in *Human Personality*, since he asks for "an intermediate conception of space—something between space as we know it in the material world and space as we imagine it to disappear in the ideal world." He suggests that in dreams and trance there may be a clearer and more complete perception of space than is at present possible to us. A corresponding sublimation of the time-sense is no less necessary to account for time in dreams. Although we seem to triumph over space and time to such a tune as to eliminate them, dream experiences have both form and sequence. Now because form presupposes space, and time is implicit in sequence, there arises the necessity for that "intermediate conception" of both space and time provided by our hypothesis.

THE PHENOMENON OF PAUSE

Let us conceive of sleep less narrowly than we are accustomed to: think of it only as one phase of the phenomenon of pause, of arrested physical activity, universal throughout nature. The cell itself experiences fatigue and goes to sleep—"perchance to dream." Modern experimental science in the domain of physiology and psychology proves that we see and do not see, hear and do not hear, feel and do not feel, in successive instants. We are asleep, in other words, not merely hour by hour, but moment by moment—and perhaps age by age as well.

Where is consciousness during these intervals, long or short, when the senses fail to respond to the stimuli of the external world? It is somewhere else, awake to some other environment. Though we may not be able to verify this from our own experience, there are methods whereby it can be verified. Clairvoyance is one of these, hypnotism is another—that kind of hypnotism whereby an entranced person is made to give a report of his excursions and adventures in the mysterious House of Sleep. It is a well-known fact that these experiences increase in intensity, coherence, and in a certain sort of omniscience, directly in proportion to the depth of the trance. The revelations obtained in this way are sometimes amazing. The inherent defect of this method of obtaining information is the possibility of deception, and for that reason science still looks askance at all evidence drawn from this source. But in essaying to write a book about the fourth dimension from any aspect but the mathematical, the author has put himself outside the pale of orthodox science; therefore he is under no compulsion to ignore a field so rich merely because it appears to be tainted by a certain amount of fallibility, and is even under suspicion of fraud. Diseased oysters, though not edible, produce pearls, and a pearl of great price is the object of this quest. Let us glance, therefore, at the findings of hypnotism and kindred phenomena.

VII

THE NIGHT SIDE OF CONSCIOUSNESS

THE FIELD OF PSYCHIC RESEARCH

IT is difficult to divest the words hypnotism and clairvoyance of certain sordid and sinister associations. We are apt to think of them as only urban flora of the dust and dark, cultivated for profit by itinerant professors and untidy sibyls. Larger knowledge of the "night side" of human nature, however, profoundly modifies this view. The invoked image is then of some hushed and studious chamber where a little group of people sit attentive to the voice of one entranced—listeners at the keyhole of the door to another world. This *news from nowhere*, garnered under so-called test conditions and faithfully recorded, has grown by now to a considerable literature, accessible to all—one with which every well-informed person is assumed to have at least a passing acquaintance.

A marked and constant characteristic of trance phenomena consists of an apparent confusion between past, present and future. As in the game of three-card monte, it appears impossible to tell in what order the three will turn up—*was, is* and *will be,* lose their special significance. Clairvoyance, in its time aspect, whether spontaneous, hypnotically induced, or self-induced, is susceptible of classification as post-vision, present vision, and pre-vision. Post-vision is that in which past events are not recollected merely, but seen or experienced. It is the past become present. Present vision is clairvoyance of things transpiring elsewhere; the present, remote in space, but not in time. Pre-vision is the future in the present. These various orders of *clear-seeing* transcend the limits of the actual knowledge and experience of the seer. This classification and these definitions are important only to us, to whom past, present, and future stand sharply differentiated in thought and in experience; not to the clairvoyant, who though bound in body to our space and time is consciously free in a world where these discriminations vanish. Why do they vanish? This question can best be answered by means of a homely analogy.

For a symbol of the flow of time in waking consciousness, imagine yourself in a railway carriage which jogs along a main-travelled line at a rate predetermined by the time-table. You approach, reach and pass such stations as are intersected by that particular railway, and you get a view of the landscape which every other traveller shares. Having once left a station, you cannot go back to it, nor can you arrive at places farther along the line before the train itself takes you there. Compare this with the freedom to do either of these things, and any number of others, if you suddenly change from the train to an automobile. Then in effect you have the freedom *of a new dimension.* In the one case, you must travel along a single line at a uniform rate; in the other, you are able to strike out in *any* direction and regulate your speed at will. You can go back to a place after the train has left it; you can go forward to some place ahead, before the train arrives, or you can strike out into and traverse new country. In short, your freedom, temporal and spatial, will be related to that of the train-bound traveller, somewhat as is trance consciousness to everyday waking life.

MODIFYING THE PAST

Modern psychology has demonstrated the existence of a great undercurrent of mental and emotional life, transcending the individual's conscious experience, in which the most complex processes are carried on without the individual's conscious participation. The clearest symbol by which this fact may be figured to the imagination is the one already presented: the comparison of the subjective field to *a plane,* in which the conscious experience of the individual is represented by *a single line.* In sleep and trance we have an augmented freedom of movement and so are able to travel here and there, backward and forward, not only among our own "dissociated memories" but in that greater and more mysterious demesne which comprehends what we call the future, as well as the present and the past.

The profound significance of the dissociation and sublimation of memory by hypnotism, or by whatever other means the train of personal experience and recollection can be thrown off the track, appears to have been ignored on its theoretical side—that is, as establishing the return of time. It has been cleverly turned to practical account, however, in the treatment of disease. By a series of painstaking and brilliant experiments, the demonstration of the rôle played by "dissociated memories" in causing certain functional nervous and mental troubles has been achieved. It has been shown that severe emotional shocks, frights, griefs, worries, may be—and frequently are—completely effaced from conscious recollection, while continuing to be vividly remembered in the depths of the subconscious. It has been shown that thence they may, and frequently do, exercise a baleful effect upon the whole organism, giving rise to disease symptoms, the particular type of which were determined by the victim's self-suggestion. As a preliminary to effecting a permanent cure to such disorders, it is necessary to get at these dissociated memories and drag them back into the full light of conscious recollection. To get at them, medical psychologists make use of hypnotism, automatic writing, crystal-gazing—in short, of any method which will force an entrance into that higher time-world through which the forgotten past may become the present. This accomplished, and the crucial moment recovered and transfixed, the victim of the aborted opportunity is led to deal with it as one may deal with the fluid, and may not deal with the fixed. Again his past is plastic to the operation of his intelligence and his will. Here is glad news for mortals: the past recoverable and in a manner revocable!

Buddha taught that all sin is ignorance, and this teaching has escaped oblivion because its truth has echoed in so many human hearts. We find that it is possible to deal with our old ignorances in the light of later knowledge. What is this but *the self-forgiveness of sins?* Subconsciously we may be always at work, mending the past. *Repentance* is the conscious recognition of some culmination of this obscure process, when the heart is suffused with the inner gladness of liberation from the payment of old karmic debts. Christ's words, "Thy sins are forgiven," spoken to the woman who washed his feet with her tears, sanctions this idea—that the past is remediable by knowledge and by love.

Conceding this much we must equally admit the possibility of moulding the future—of adjusting the will to the event which shall befall. If the present moment can again intersect the stream of past conscious experience, it may equally do so with regard to the future. This brings up the tremendous questions of free will and fore-ordination. Upon these the Oriental doctrines of karma and reincarnation cast the only light by which the reason consents to be guided. As these doctrines are intimately related both to higher time and to trance revelations, some consideration of karma and reincarnation may appropriately find place here.

KARMA AND REINCARNATION

Karma is that self-adjusting force in human affairs which restores harmony disturbed by action. It is the moral law of compensation, and by its operation produces all conditions of life, misery and happiness, birth, death, and re-birth; itself being both the cause and the effect of action. Its operation is indicated in the phrase, "Whatsoever a man soweth, that shall he also reap."

The essential idea of reincarnation is indicated in the following quotation from the *Upanishads:* "And as a goldsmith, taking a piece of gold, turns it into another, newer, and more beautiful shape, so does this Self, having thrown off this body and dispelled all ignorance, make unto himself another and more beautiful shape."

Reincarnation is the periodic "dip" of an immortal individual into materiality for the working out of karma, after an interval, long or short, spent under other conditions of existence. These alternations constitute the broader and deeper diapason of human life, of which the change from waking to sleeping represents the lesser, and the momentary awareness and unawareness of the sense mechanism to stimulation, the least.

Thus a physical incarnation, in the broadest sense of the term, is the interval—long or short—of the immersion of consciousness in materiality. Under fatigue the cell life withdraws; that is it ceases to respond to physical stimuli and so passes out of incarnation. When this occurs *en masse* there transpires that hiatus of the personal consciousness called sleep, and while sleep lasts the personality is out of incarnation. After death—in the interval between one life and the next—the specific memories of the personality fade out as in sleep, or rather become latent, leaving the soul, the permanent life-center, clear and colorless, a mysterious focus of spiritual forces and affinities (the seeds of karma) ready for another sowing in the world of men. This center of consciousness is thereupon drawn to the newly forming body, the life environment of which will rightly and justly—perhaps retributively—bring the tendencies and characteristics of the conscious center into objectivity again. Character is destiny, and character is self-created—"All that we are is the result of what we have thought." But in the vast complexity and volume of human life there is a constant production of forms, with all the varieties of characteristics and capacities requisite to meet the needs of every soul, thirsty for the destiny that awaits it; and here heredity plays its part—the parent provides not the mind nor the soul, but only the *vehicle of expression* of another karmically related soul. Beyond the individual soul is the world-soul, which periodically incarnates in the humanity of a planet, and beyond the worlds of a single system, *sun-souls* incarnate in suns and congeries of suns. And beyond all, and within all—*One!*

The profound and pregnant doctrines of karma and reincarnation, here so sketchily outlined, are but expansions of one of the fundamental propositions of all Eastern philosophical systems, that the effect is the unfoldment of the cause in time.

To omit a consideration of karma and reincarnation in connection with higher time would be to force a passage and then not follow where it leads. The idea of time curvature is implicit in the ideas of karma and reincarnation. For what is karma but *the return of time*, the flowering in the present of some seed sown elsewhere and long ago? And what is reincarnation but the major cycle of that sweep into objective existence and out again, of which the alternation between waking and sleeping is the lesser counterpart?

COLONEL DE ROCHAS' EXPERIMENTS

During the past few years evidence has been accumulating that we never really forget anything. We have rediscovered the memory of the subconscious mind. It is generally known that in the mesmeric or somnambulistic sleep things hopelessly beyond recall for the habitual mind come to the surface in fragments or in whole series, as the case may be. It is perhaps news to some readers however that the memory of past lives has been recovered in this way. Thus is confirmed the Eastern secret teaching that could we remember our dream experiences we should recover the knowledge of our past incarnations.

Among the achievements of Eastern hypnotism is the recovery of the memory of past births. Colonel de Rochas appears to have paralleled this achievement in the West. Certain of his experiments have been admirably reported by Maurice Maeterlinck in the eighth chapter of *Our Eternity*. Maeterlinck's account, somewhat condensed, is given here because it so well illustrates the liberation of consciousness from the tyranny of time as we conceive it. He says:

"First of all, it is only right to say that Colonel de Rochas is a savant who seeks nothing but objective truth and does so with a scientific strictness and integrity that have never been questioned. He puts certain exceptional subjects into a hypnotic sleep and, by means of downward passes, makes them trace back the whole course of their existence. He thus takes them successively to their youth, their adolescence and down to the extreme limits of their childhood. At each of these hypnotic stages, the subject reassumes the consciousness, the character and the state of mind which he possessed at the corresponding stage in his life. He goes over the same events, with their joys and their sorrows. If he has been ill, he once more passes through his illness, his convalescence and his rccovery.

"Let us, to come to details, take one of the simplest cases. The subject is a girl of eighteen, called Joséphine. She lives at Voiron, in the department of Isère. By means of downward passes she is brought back to the condition of a baby at its mother's breast. The passes continue and the wonder-tale runs its course. Joséphine can no longer speak; and we have the great silence of infancy, which seems to be followed by a silence more mysterious still. Joséphine no longer answers except by signs: *she is not yet born.* 'She is floating in darkness.' They persist; the sleep becomes heavier; and suddenly, from the depths of that sleep, rises the voice of another being, a voice unexpected and unknown, the voice of a churlish, distrustful and discontented old man. They question him. At first he refuses to answer, saying that 'of course he's there, and he's speaking'; that 'he sees nothing'; and 'he's in the dark.' They increase the number of passes and gradually gain his confidence. His name is Jean Claude Bourdon; he is an old man; he has long been ailing and bed-ridden. He tells the story of his life. He was born at Champvent, in the parish of Polliat, in 1812. He went to school until he was eighteen and served his time in the army with the Seventh Artillery at Besançon; and he describes his gay time there, while the sleeping girl makes gestures of twirling an imaginary moustache. When he goes back to his native place, he does not marry, but he has a mistress. He leads a solitary life (I omit all but the essential facts), and dies at the age of seventy, after a long illness.

"We now hear the dead man speak; and his posthumous revelations are not sensational, which, however, is not an adequate reason for doubting their genuineness. He feels himself growing out of his body; but he remains attached to it for a fairly long time. His fluidic body, which is at first diffused, takes a more concentrated form. He lives in darkness, which he finds disagreeable; but he does not suffer. At last, the night in which he is plunged is streaked with a few flashes of light. The idea comes to reincarnate himself and he draws near to her who is to be his mother (that is, the mother of Joséphine). He encircles her until the child is born, whereupon he gradually enters the child's body. Until about the seventh year, his body is surrounded by a sort of floating mist, in which he used to see many things which he has not seen since.

"The next thing to be done is to go back beyond Jean Claude. A mesmerization lasting nearly three-quarters of an hour, without lingering at any intermediate stage, brings the old man back to babyhood. A fresh silence, a new limbo; and then, suddenly, another voice and an unexpected individual. This time it is an old woman who has been very wicked; and so she is in great torment (she is dead, at the actual instant; for, in this inverted world, lives go backward and of course begin at the end). She is in deep darkness, surrounded by evil spirits. She speaks at first in a faint voice, but always gives definite replies to the questions put to her, instead of cavilling at every moment, as Jean Claude did. Her name is Philoméne Carteron.

" 'By intensifying the sleep,' adds Colonel de Rochas, whom I will now quote, 'I induce the manifestations of a living Philoméne. She no longer suffers, seems very calm and always answers coldly and distinctly. She knows that she is unpopular in the neighbourhood, but no one is a penny the worse and she will be even with them yet. She was born in

1702; her maiden name was Philoméne Cherpigny; her grandfather on the mother's side was called Pierre Machon and lived in Ozan. In 1732 she married, at Chevroux, a man named Carteron, by whom she had two children, both of whom she lost.' "

Before her incarnation, Philoméne had been a little girl who died in infancy. Previous to that, she was a man who committed murder, and it was to expiate this crime that she endured such suffering in the darkness, and after her life as a little girl, when she had no time to do wrong. Colonel de Rochas did not think it wise to carry the hypnosis further, because the subject appeared exhausted and her paroxysms were painful to watch. He obtained analogous and even more surprising results with other subjects.

Maeterlinck's comments upon all this are of negligible value. He pays a fine tribute to the theory of reincarnation: "There was never a more beautiful, a juster, a purer, a more moral, fruitful and probable creed," he says; yet for all that, it is clear that he has not been at pains fully to inform himself of the Eastern teaching, and that its deeper meanings escape him.

Colonel de Rochas' success, and that of all other experimenters along these lines, is due to their unconscious following of the Eastern method. He himself says that he "avoided everything that should put the subject on a definite tack"—that is, he refrained from voluntary suggestion.

Having referred so frequently and so familiarly to the Eastern belief in reincarnation, and hinted at a more solid foundation for that belief than the single series of experiments above referred to, it would be unfair to the reader not to gratify his curiosity more fully in regard to these matters. In the light of our hypothesis they take on an importance which justifies their further consideration here.

VIII

THE EASTERN TEACHING

ORIENTAL PHYSICS AND METAPHYSICS

WESTERN physical science, pursued with ardor and devotion for the past hundred years, has attained to a control over physical phenomena little short of magical, but in our understanding and mastery of subjective phenomena we are far behind those Eastern peoples who have made these matters the subject of study and experiment for thousands of years. The informed Hindu, rightly or wrongly, regards the Western practice of hypnotism both in its methods and in its results with mingled horror and contempt. To him it is not different from Black Magic, pernicious to operator and subject alike, since it involves an unwarrantable tyranny of the will on the part of the operator, and a dangerous submission to the obsession of an invading will on the part of the subject. Eastern hypnotism—at its highest and best—is profoundly different from Western, in that the sanctity of the individual is respected. Its aim is not to enslave the will, but temporarily, and under favourable conditions, to emancipate consciousness from its physical limitation.

Eastern practical psychology and metaphysics can be understood only through a knowledge of Eastern physics. These we would call *transcendental,* since they recognize not one theatre of consciousness but three: the gross, the subtle, and the pure. These correspond to the material, the etherial, and the empyreal worlds of Greek philosophy, and to the physical, astral, and mental planes of modern Theosophy. They may be thought of as universal substance in three different octaves of vibration or as three "vehicles" or instruments differently attuned. Upon them the trained will of man is able to act directly, for the reason that—as claimed by Balzac—it is a *living* force.

In Eastern hypnotism the gross vibrations of the physical vehicle are inhibited by the will of the operator, putting the body of the subject to sleep, whereat the consciousness, free in its subtle body, awakens to a dimensionally higher world. The operator, by means of questions, reaps such profit as he may by following the *true dreams* of the entranced subject, scrupulously refraining from imposing his own will further than is necessary to obtain the information which he seeks. The higher power of Eastern hypnotism, totally unknown in the West, consists of inhibiting the subtle vibrations of the astral vehicle also, permitting the consciousness to revert to its "pure" condition. In these deep states of trance the subject is able to communicate knowledges shut away from the generality of men—among them the knowledge of past births.

THE SELF-RECOVERED MEMORY OF PAST BIRTHS

The strength of will necessary to accomplish this higher power of hypnotism is achieved by arduous and long-continued exercises in concentration, by the practice of a strict though unconventional morality, and by submission to a physical regimen which few Occidentals would care to undergo. Severe as is this training, it is less so than that which the true Yogi imposes upon himself, and its fruits are less. The achievement to which he addresses himself is far beyond that of the most accomplished hypnotist or miracle-working fakir, for the Yogi scorns all supernormal powers, even while possessing them. The Yogi, as the word implies—it means literally union—seeks to unite himself with his own higher self, the eternal and immortal part of his own nature, and the achievement of this brings with it the freedom of the three worlds at all times, and in full consciousness. As this involves an inward turning of the mind and will, and the withdrawal from the ordinary active life of humanity, he alone is witness of his own success.

The knowledge of past births which may be

obtained by the questionable and cumbersome method of hypnotism is one of the wayside flowers which the Yogi may pluck, if he will, on his path toward perfection. There are definite rules for the attainment of this knowledge, and they conform so closely to Colonel de Rochas' method—save for the fact that operator and subject are one and not twain—that it will be interesting to give them here. The ensuing passage is from the *Vishuddhi Marga*, or *Path of Purity*, a work written some sixteen hundred years ago by the famous sage, Buddhaghosha, whose name signifies the Voice of Buddha, the revealer of Buddha's teachings. It is quoted in Charles Johnston's *The Memory of Past Births*.

"The devotee, then, who tries for the first time to call to mind former states of existence, should choose a time after breakfast, when he has returned from collecting alms, and is alone and plunged in meditation, and has been absorbed in the four trances in succession. On rising from the fourth trance, which leads to the higher powers, he should consider the event which last took place, namely, his sitting down; next, the spreading of the mat; the entering of the room; the putting away of bowl and robe; his eating; his leaving the village; his going the rounds of the village for alms; his entering the village for alms; his departure from the monastery; his offering adoration in the courts of the shrine and of the Bodhi tree;

his washing the bowl; what he did between taking the bowl and rinsing his mouth; what he did at dawn; what he did in the middle watch of the night; what he did in the first watch of the night. Thus he must consider what he did for a whole day and night, going backwards over it in reverse order.

"In the same reverse order he must consider what he did the day before, the day before that, up to the fifth day, the tenth day, a fortnight ago, a month ago, a year ago; and having in the same manner considered the previous ten and twenty years, and so on up to the time of his conception in this birth, he must then consider the name and form which he had at the moment of death in his last birth. But since the name and form of the last birth came quite to an end, and were replaced by others, this point of time is like thick darkness, and difficult to be made out by the mind of any person still deluded. But even such a one should not despair nor say: 'I shall never be able to penetrate beyond conception, or take as the object of my thought the name and form which I had in my last birth, at the moment of death,' but he should again and again enter the trance which leads to the higher powers, and each time he rises from the trance, he should again intend his mind upon that point of time.

"Just as a strong man in cutting down a mighty tree to be used as the peaked roof of a pagoda, if the edge of his axe be turned in lopping off the branches and twigs, will not despair of cutting down the tree, but will go to an iron-worker's shop, have his axe sharpened, return, and go on with his cutting; and if the edge of his axe be turned a second time, he will a second time have it sharpened, and return, and go on with his cutting; and since nothing that he chopped once needs to be chopped again, he will in no long time, when there is nothing left to chop, fell that mighty tree. In the same way the devotee rising from the trance which leads to the higher powers, without considering what he has considered once, and considering only the moment of conception, in no long time will penetrate beyond the moment of conception, and take as his object the name and form which he had at the moment of death, in his last birth.

"His alert attention having become possessed of this knowledge, he can call to mind many former states of existence, as, one birth, two births, three births, four births, five births, and so on, in the words of the text."

This quotation casts an interesting light upon Eastern monasticism. The Buddhist monasteries are here revealed as schools of practical psychology, the life of the monk a life of arduous and unceasing labor, but labor of a sort which masks as idleness. The successive "initiations" which are the milestones on that "Path of Perfection" upon which the devotee has set his feet represent successive emancipations of consciousness gained through work and knowledge. Their nature may best be understood by means of a fanciful analogy.

RELEASE

If we assume that all life is conscious life, as much aware of its environment as the freedom of movement of its life vehicle in that environment permits, a corpuscle vibrating in a solid would have a certain sense of space and of movement in space gained from its own experience. Now imagine the solid, which is its world, to be subjected to the influence of heat. When the temperature reached a certain point the solid would transform itself into a liquid. To the corpuscle all the old barriers would seem to be broken down; space would be different, time would be different, and its world a different place. Again, at another increase of temperature, when the liquid became a gas, the corpuscle would experience a further emancipation: it would possess a further freedom, with all the facts of its universe to learn anew.

Each of these successive crises would constitute for it *an initiation,* and since the heat has acted upon it from within, causing an expansion of its life vehicle, it would seem to itself to have attained to these new freedoms through self-development.

The parallel is now plain to the reader. The corpuscle is the Yogi, bent on liberation; the heat which warms him is the Divine Love, centred in his heart; his initiations are the successive emancipations into higher and higher spaces, until he attains Nirvana—inherits the kingdom prepared for him from the foundation of the world. As latent heat resides in the corpuscle, so is *Release* hidden in the heart—release from time and space. The perception of this prompted the exultant apostrophe of Buddha, "Looking for the maker of this tabernacle, I have run through a course of many births, not finding him; and painful is birth again and again. But now, maker of the tabernacle, thou hast been seen; thou shalt not make up this tabernacle again. All thy rafters are broken, thy ridgepole is sundered; the mind, approaching the Eternal, has attained the extinction of all desires."

Upon the mystery of Nirvana the Higher Space Hypothesis casts not a little light. To "approach the Eternal" can only be to approach a condition where time is not. Because there is an escape from time in proportion as space dimensions are added to, and assimilated by, consciousness, any development involving this element of space conquest (and evolution is itself such a development) involves time annihilation also. To be in a state of desire is to be conditioned by a limitation, because one can desire only that which one has not or is not. The extinction of a desire is only another name for the transcending of a limitation—of all desires, of all limitations. If these limitations are of space they are of time also; therefore is the "approach to the Eternal" through the "extinction of all desire." Christ said, "Him that overcometh will I make a pillar of the temple of my God, and he shall go no more out"—go out, that is, into incarnation—into "time, besprent with seven-hued circumstance."

Such are the testimonies of the world-saviors regarding the means and end of liberation. Below them on the evolutionary ladder stand the mystics, earth-bound, but soul-free; below them, in turn, yet far above common humanity, stand the men of genius, caught still in the net of passion, but able in their work to reflect something of the glory of the supernal world. Let us consider in the next two chapters each of these in turn.

IX

THE MYSTICS

HERMES TRISMEGISTUS

THE mystic however far removed he may be from Nietzsche's ideal of the Superman, nevertheless represents superhumanity *in the domain of consciousness.* By means of quotations, taken almost at random from the rich literature of mysticism, the author will attempt to show that the consciousness of the mystic involves the awareness of dimensionally higher worlds. The first group of quotations is culled from certain of the sacred books of Hermes Trismegistus.

"Comprehend clearly" (says Hermes to Asclepios) *"that this sensible world is enfolded, as in a garment, by the supernal world."*

We think of our three-dimensional space, "the sensible world," as *immersed* in higher space; "enfolded as in a garment," therefore. And we think of the objects of our world as having extension in a dimensionally higher region, that "supernal world" in which the phenomena of this sensible world arise. For:

"Celestial order reigns over terrestrial order: all that is done and said upon earth has its origin in the heights, from which all essences are dispensed with measure and equilibrium: nor is there anything which does not emanate from one above and return thither."

THE PAGE AND THE PRESS

The idea of an all-embracing unity within and behind the seeming manifoldness of life forms the ground rhythm of all inspired literature, sacred and profane alike. For clarity and conciseness it would be difficult to improve upon the formulation of this idea contained in the following fragment:

"In the manifold unity of universal life the innumerable individualities distinguished by their variations are, nevertheless, united in such a manner that the whole is one, and that everything proceeds from unity.

"For all things depend upon unity, or develop from it, and because they appear distant from one another it is believed that they are many, whereas in their collectivity they form but one."

Now nothing so successfully resolves this paradox of the one and the many as the concept that the things of this world are embraced and united in a dimensionally higher world in a manner analogous to that in which all conic sections are embraced and united within the cone. A more elaborate analogy may serve to make this clearer to the mind.

Conceive of this printed page as a plane-world in which every letter is a person; every word a family; and phrases and sentences are larger communities and groups. These "innumerable individualities, distinguished by their variations" must needs seem to themselves as "distant from one another," their very differences of form and arrangement a barrier to any superior unity. Yet all the while, solely by reason of this diversity, they are coöperating towards an end of which they cannot be aware. The mind of the reader unites and interprets the letters into continuous thought, though they be voiceless as stones to one another. Even so may our sad and stony identities spell out a world's word which we know not of, by reason of our singularity and isolation. Moreover, in the electrotype block, the solid of which the printed page constitutes a plane presentment, all the letters are actually "united in such a manner that the whole is one." The metal that has moulded each into its significant form amalgamates them into a higher unity. So also the power that makes us separate is the same power that makes us one.

THE SHIP AND ITS CAPTAIN

Here follows the lament of the souls awaiting incarnation:

"Behold the sad future in store for us—to minister to the wants of a fluctuating and dissoluble body! No more may our eyes distinguish the souls divine! Hardly through these watery spheres shall we perceive, with sighs, our ancestral heaven: at intervals even we shall cease altogether to behold it. By this disastrous sentence direct vision is denied to us; we can see only by the aid of the outer light; these are but windows that we possess—not eyes. Nor will our pain be less when we hear in the fraternal breathing of the winds with which no longer can we mingle our own, since ours will have for its dwelling, instead of the sublime and open world, the narrow prison of the breast!"

That the soul—the so-called subliminal self—draws from a broader, deeper experience than the purely rational consciousness is a commonplace of modern psychology. Hinton conceives of the soul as *higher-dimensional* with relation to the body, but so concerned with the management and direction of its lower-dimensional vehicle as to have lost, for the time being, its orientation, thinking and moving only in those ways of which the body is capable. The analogy he uses, of a ship and its captain, is so happy, and the whole passage has so direct a bearing upon the Hermetic fragment quoted, that it is given here entire.

"I adopt the hypothesis that that which thinks in us has an ample experience, of which the intuitions we use in dealing with the world of real objects are a part; of which experience, the intuition of four-dimensional forms and motions is also a part. The process we are engaged in intellectually is the reading of the obscure signals of our nerves into a world of reality, by means of intuitions derived from the inner experience.

"The image I form is as follows: Imagine the captain of a modern battleship directing its course. He has his charts before him; he is in communication with his associates and subordinates; can convey his messages and commands to every part of the ship, and receive information from the conning tower and the engine room. Now suppose the captain, immersed in the problem of the navigation of his ship over the ocean, to have so absorbed himself in the problem of the direction of the craft over the plane surface of the sea that he forgets himself. All that occupies his attention is the kind of movement that his ship makes. The operations by which that movement is produced have sunk below the threshold of his consciousness; his own actions, by which he pushes the buttons, gives the orders, are so familiar as to be automatic; his mind is on the motion of the ship as a whole. In such a case we can imagine that he identifies himself with the ship; all that enters his conscious thought is the direction of its movement over the plane surface of the ocean.

"Such is the relation, as I imagine it, of the soul to the body. A relation which we can imagine as existing momentarily in the case of the captain is the normal one in the case of the soul with its craft. As the captain is capable of a kind of movement, an amplitude of motion, which does not enter into his thoughts with regard to the directing of the ship over the plane surface of the ocean, so the soul is capable of a kind of movement, has an amplitude of motion, which is not used in its task of directing the body in the three-dimensional region in which the body's activity lies. If for any reason it becomes necessary for the captain to consider three-dimensional motions with regard to his ship, it would not be difficult for him to gain the materials for thinking about such motions; all he has to do is to call experience into play. As far as the navigation of the ship is concerned, however, he is not obliged to call on such experience. The ship as a whole simply moves on a surface. The problem of three-dimensional movement does not ordinarily concern its steering. And thus with regard to ourselves all those movements and activities which characterize our bodily organs are three-dimensional; we never need to consider the ampler movements. But we do more than use these movements of our body to effect our aims by direct means; we have now come to the pass when we act indirectly on nature, when we call processes into play which lie beyond the reach of any explanation we can give by the kind of thought which has been sufficient for the steering of our craft as a whole.

"When we come to the problem of what goes on in the minute and apply ourselves to the mechanism of the minute, we find our habitual conceptions inadequate. The captain in us must wake up to his own intimate nature, realize those functions of movement which are his own, and in the virtue of his knowledge of them apprehend how to deal with the problems he has come to."

—*The Fourth Dimension.*

How more accurately and eloquently could "the captain in us," momentarily aroused, give voice to his predicament, than in the words, "*Instead of the sublime and open world, the narrow prison of the breast.*"

DIRECT VISION

The "watery spheres" in the Hermetic fragment are of course the eyes, a mechanism inferior in many ways to the camera of man's own devising. The phenomena of clairvoyance make known a mode of vision which is confined to no specific sense organ, approximating much more closely to true perception than does physical sight. Mr. C. W. Leadbeater in *Clairvoyance* specifically affirms that this higher power of sight is four-dimensional. He says: "The idea of the fourth dimension as expounded by Mr. Hinton is the only one which gives any kind of explanation down here of astral vision . . . which lays every point in the interior of a solid body absolutely open to the gaze of the seer, just as every point of the interior of a circle lies open to the gaze of a man looking down upon it." "I can see all around and every way," exclaims one of the psychometers reported in William Denton's *The Soul of Things*.

The "outer light" by which the physical eye is able to see objects is sunlight. Upon this clairvoyant vision in no wise depends, involving, as it does, other octaves of vibration. We should be able to receive ideas of this order without incredulity since the advent of "dark" photography and the ultra-violet microscope. By aid of the latter, photographs are taken in absolute darkness, the lenses used being transparent to light rays to which the eye is insensible, but which are active photographically.

The foregoing passages from *The Virgin of the World* show a remarkable resemblance between the Hermetic philosophy and modern higher-space thought. The parallelism is not less striking in the case of certain other mystic philosophers of the East.

PLATO'S SHADOW-WATCHERS

"Parmenides," says Hinton, "and the Asiatic thinkers with whom he is in close affinity, propound a theory of existence which is in close accord with a conception of a possible relation between a higher and a lower-dimensional space." He concludes, "Either one of two things must be true, that four-dimensional conceptions give a wonderful power of representing the thought of the East, or that the thinkers of the East must have been looking at and regarding four-dimensional existence."

It would not be difficult to re-state, in terms of our hypothesis, Plato's doctrine of an enduring archetypal world of ideas reflected in a world of transitory images and appearances. Fortunately, Plato has relieved the author of that necessity by doing it himself in his wonderful allegory of the shadow-watchers in *The Republic.* The trend of his argument is clear: as its shadow is to a solid object, so is the object itself to its archetypal idea. This is the manner in which he presents this thought:

"Imagine a number of men living in an underground cavernous chamber, with an entrance open to the light, extending along the entire length of the cavern, in which they have been confined, from their childhood, with their legs and neck so shackled, that they are obliged to sit still and look straight forwards, because their chains render it impossible for them to turn their heads round: and imagine a bright fire burning some way off, above and behind them, and an elevated roadway passing between the fire and the prisoners, with a low wall built along it, like the screens which conjurors put up in front of their audience, and above which they exhibit their wonders."

"I have it," he replied.

"Also, figure to yourself a number of persons walking behind this wall, and carrying with them statues of men, and images of other animals, wrought in wood, stone, and all kinds of materials, together with various other articles, which overtop the wall; and, as you might expect, let some of the passers-by be talking, and others silent."

"You are describing a strange scene, and strange prisoners."

"They resemble us," I replied. "For let me ask you, in the first place, whether persons so confined could have seen anything of themselves or of each other, beyond the shadows thrown by the fire upon the part of the cavern facing them."

"Certainly not, if you suppose them to have been compelled all their lifetime to keep their heads unmoved."

"And is not their knowledge of the things carried past them equally limited?"

"Unquestionably it is."

"And if they were able to converse with one another, do you not think that they would be in the habit of giving names to the objects which they saw before them?"

"Doubtless they would."

"Again: if their prison house returned an echo from the part facing them, whenever one of the passers-by opened his lips, to what, let me ask you, could they refer the voice, if not to the shadow which was passing?"

"Unquestionably they would refer it to that."

"Then surely such persons would hold the shadows of the manufactured articles to be the only realities."

"Without a doubt they would."

Plato (in the person of Socrates) then considers what would happen if the course of nature brought to the prisoners a release from their fetters and a remedy for their foolishness, and concludes as follows:

"Now this imaginary case, my dear Glaucon, you must apply in all its parts to our former statements, by comparing the region which the eye reveals, to the prison-house, and the light of the fire therein to the power of the sun; and if, by the upward ascent and the contemplation of the upper world, you understand the mounting of the soul in the intellectual region, you will hit the tendency of my own surmises . . . the view which I take of the subject is to the following effect."

Briefly, the view taken is that the "Form of Good" perceived by the mind is the source of everything that is perceived by the senses. This is equivalent to saying that the objects of our three-space world are projections of higher-dimensional realities—that there is a supernal world related to this world as a body is related to the shadow which it casts.

SWEDENBORG

Emerson, in his *Representative Men,* chose Swedenborg as the representative mystic. He accepted Swedenborg's way of looking at the world as universally characteristic of the mystical temperament. The Higher Space Theory was unheard of in Swedenborg's day, nevertheless in his religious writings—thick clouds shot with lightning—the idea is implicit and sometimes even expressed, though in a terminology all his own.

To Swedenborg's vision, as to Plato's, this physical world is a world of ultimates, in all things correspondent to the causal world, which he names "heaven." *"It is to be observed,"* he says, *"that the natural world exists and subsists from the spiritual world, just as an effect exists from its efficient cause."*

According to Swedenborg, conditions in "heaven" are different from those in the world: space is different, distance is different. He says, "*Space in heaven is not like space in the world, for space in the world is fixed, and therefore measurable: but in heaven it is not fixed and therefore cannot be measured.*"

Herein is suggested a *fluidic* condition, singularly in accord with certain modern conceptions in theoretical physics. Commenting upon the significance of Lobatchewsky's and Bolyai's work along the lines of non-Euclidian geometry, Hinton says, "By immersing the conception of distance in matter, to which it properly belongs, it promises to be of the greatest aid in analysis, for the effective distance of any two particles is the result of complex material conditions, and cannot be measured by hard and fast rules."

The higher correlative of physical distance is a difference of state or condition, according to the Swedish seer. "*Those are far apart who differ much,*" he says "*and those are near who differ little.*" Distance in the spiritual world, he declares, originates solely "*in the difference in the state of their minds, and in the heavenly world, from the difference in the state of their loves.*" This immediately suggests the Oriental teaching that the place and human environment into which a man is born have been determined by his own thoughts, desires, and affections in anterior existences, and that instant by instant all are determining their future births. The reader to whom the idea of reincarnation is repellent or unfamiliar may not be prepared to go this length, but he must at least grant that in the span of a single lifetime thought and desire determine action, and consequently, position in space. The ambitious man goes from the village to the city; the lover of nature seeks the wilds; the misanthrope avoids his fellowmen, the gregarious man gravitates to crowds. We seek out those whom we love, we avoid those whom we dislike; everywhere the forces of attraction and repulsion play their part in determining the tangled orbits of our every-day lives. In other words, the subjective, and (hypothetically) higher activity in every man records itself in a world of three dimensions as action upon an environment. Thought expresses itself in action, represents itself in form and so "externalizes" itself in space.

Observe how perfectly this fits in with Swedenborg's contention that physical remoteness has for its higher correspondence a difference of love and of interest; and physical juxtaposition a similarity of these. In heaven, he says, "Angels of similar character are as it were spontaneously drawn together." So would it be on earth, but for impediments inherent in our terrestial space. Swedenborg's angels are men freed from these limitations. We suffer because the free thing in us is hampered by the restrictions of a space to which it is not native. Reason sufficient for such restriction is apparent in the success that crowns every effort at the annihilation of space, and the augmentation of power and knowledge that such effort brings. It would appear that a narrowing of interest and endeavor is always the price of efficiency. The angel is confined to "the narrow prison of the breast" that it may react upon matter, just as an axe is narrowed to an edge that it may cleave.

MAN THE SPACE-EATER

Man has been called the thinking animal. *Space-eater* would be a more appropriate title, since he so dauntlessly and persistently addresses himself to overcoming the limitations of his space. To realize his success in this, compare for example the voyage of Columbus' caravels with that of an ocean liner; or travelling by stage coach with *train de luxe.* Consider the telephone, the phonograph, the cinematograph, from the standpoint of space-conquest—and wireless telegraphy which sends forth messages in every direction, over sea and land. Most impressive of all are the achievements in the domain of astronomy. One by one the sky has yielded its amazing secrets, till the mind roams free among the stars. The reason why there are today so many men braving death in the air is because the conquest of the third dimension is the task to which the Zeit-Geist has for the moment addressed itself, and these intrepid aviators are its chosen instruments—sacrificial pawns in the dimension-gaining game.

All these things are only the outward and visible signs of the angel, incarnate in a world of three dimensions, striving to realize higher-spatial, or heavenly, conditions. The spectacle for example of a millionaire hurled across a continent in a special train to be present at the bedside of a stricken dear one may be interpreted as the endeavor of an incarnate soul to achieve, with the aid of human ingenuity applied to space annihilation, that which discarnate it could compass without delay or effort.

THE WITHIN AND WITHOUT

In Swedenborg's heaven "*all communicate by the extension of the sphere which goes forth from the life of every one. The sphere of their life is the sphere of their affections of love and hate.*"

This is as fair a description of thought-transference and its necessary condition as could well be devised, for as in wireless telegraphy, its mechanical counterpart, it depends upon synchronism of vibration in a "sphere which goes forth from the life of every one." Thought-transference and kindred phenomena, in which all categories of space and time lose their significance, baffle our understanding because they appear to involve the idea of being in two places

—in many places—at once, a thing manifestly at variance with our own conscious experience. It is as though the pen-point should suddenly become the sheet of paper. But strange as are these matters and mysterious as are their method, no other hypothesis so well explains them as that they are higher-dimensional experiences of the self. We have the universal testimony of all mystics that the attainment of mystical consciousness is by inward contemplation—turning the mind back upon itself. Swedenborg says, "It *can in no case be said that heaven is outside of any one, but it is within him, for every angel participates in the heaven around him by virtue of the heaven which is within him.*" Christ said, "*The Kingdom of Heaven is within you,*" and there is a saying attributed to Him to the effect that "*When the outside becomes the inside, then the Kingdom of Heaven is come.*" These and such arcane sayings as "*Know Thyself,*" engraved upon the lintels of ancient temples of initiation, powerfully suggest the possibility that by penetrating to the center of our individual consciousness we expand outwardly into the cosmic consciousness as though *in* and *out* were the positive and negative of a new dimension. By exerting a force in the negative direction upon a slender column of water in a hydraulic press, it is possible to raise in the positive direction a vast bulk of water with which that column, through the mechanism of the press, is connected. This is because both columns, the little and the big, enclose one body of fluid. The attainment of higher states of consciousness is potential in every one, for the reason that the consciousness of a greater being flows through each individual.

INTUITION AND REASON

There is the utmost unanimity in the testimony of the mystics that the world without and the world within are but different aspects of the same reality—"*The eye with which I see God is the same eye with which He sees me.*" They never weary of the telling of the solidarity and invisible continuity of life, of the inclusion not only of the minute in the vast, but of the vast in the minute. We may accept this form of perception as characteristic of consciousness in its free state. Its instrument is *the intuition,* which divines relations between diverse things through a perception of unity. The instrument of the purely mundane consciousness, on the other hand, is *the reason,* which dissevers and dissects phenomena, divining unity through correlation. Now if physical phenomena, in all their manifoldness, are lower-dimensional projections, upon a lower dimensional space, of a higher unity, then reason and intuition are seen to be two modes of one intelligence engaged in apprehending life from below (by means of the reason) through its diversity, and from above (by means of the intuition) through its unity.

Those who recognize in the intuition a valid organ of knowledge, are disposed to exalt it above the reason, but at our present state of evolution and given our environment it would seem that the reason is the more generally useful faculty of the two. In that unfolding, that manifesting of the higher in the lower—which is the idea the fourth-dimensionalist has of the world—the painstaking, minute, methodical action of the reasoning mind applied to phenomena achieves results impossible to Pisgah-sighted intuition. The power, peculiar to the reason, of isolating part after part from the whole to which it belongs, and considering them thus isolated, makes possible in the end a synthesis in which the whole is not merely glimpsed, but known to the last detail.

The method of the reason is symbolized in so trifling a thing as the dealing out one by one of a pack of cards and their reassembling. The pack has been made to show forth its contents by a process of disruption—of slicing. Similarly, if a scientist wants to gain a thorough comprehension of a complicated organism, he dissects it, or submits it to a process of slicing, studying each slice separately under the microscope while keeping constantly in mind the relation of one slice to another. This amounts to nothing less than reducing a thing from three dimensions to two, in order to know it thoroughly. Now the flux of things corresponds to the four-dimensional aspect of the world, and with this the reason finds it impossible to deal. As Bergson has so well shown, the reason cuts life into countless cross-sections: a thing must be dead before it can be dissected. This is why the higher-dimensional aspect of life, divined by the intuition, escapes rational analysis.

THE COIL OF LIFE

Swedenborg's description of "the ascent and descent of forms" and the "forces and powers" which flow therefrom, suggests, by reason of the increasing amplitude and variety of form and motion, a progression from space to space. This description is too long and involved to find place here, but its conclusion is as follows:

"*Such now is the ascent and descent of forms or substances in the greatest, and in our least universe: similar also is the descent of all forces and powers which flow from them. But all their perfection consists in the possibility and virtue of varying themselves, or of changing states, which possibility increases with their elevations, so that in number it exceeds all the series of calculations unfolded by human minds, and still inwardly involved by them:*

which infinities finally become what is finite in the Supreme. Our ideas are merely progressions by variations of form, and thus by actual changes of state."

His sense of the beauty and orderliness of the whole process, and his despair of communicating it, find characteristic utterance in the following passage:

"*If thou could'st discern, my beloved, how distinctly and ordinately these forms are arranged and connected with each other, from the mere aspect and infinity of so many wonderful things connected with each other, from the mere aspect and infinity of so many wonderful things conspiring into one, thou would'st fall down, from an inmost impulse, with sacred astonishment, and at the same time pious joy, to perform an act of worship and of love before such an architect.*"

In his description of the manner in which these forms cohere and successively unfold he introduces one of the basic concepts of higher space thought: that in the "descent of forms" from space to space, that which in the higher exists all together—that is, *simultaneously*, can only manifest itself in the lower piecemeal—that is, *successively*. He says:

"*Nothing is together in any texture or effect which was not successively introduced; and every-*

thing is therein, according as order itself introduces it: wherefore simultaneous order derives its birth, nature and perfection from successive orders, and the former is only rendered perspicuous and plain by the latter. . . . What is supreme in things successive takes the inmost place in things simultaneous: thus things superior in order super-involve things inferior and wrap them together, that these latter may become exterior in the same order: by this method first principles, which are also called simple, unfold themselves, and involve themselves in things posterior or compound: wherefore every perfection of what is outermost flows forth from inmost principles by their series: hence thy beauty, my daughter, the only parent of which is order itself."

This passage, like a proffered dish full of rare fruit, tempts the metaphysical appetite by the wealth and variety of its appeal; but not to weary the reader, the author will content himself by the abstraction of a single plum. The plum in question is simply this (and the reader is asked to read the quotation carefully again): may not every act, incident, circumstance in a human life be the "uncoiling" of a karmic aggregate? This coil of life may be thought of most conveniently in this connection as the *character* of the person, a character built up, or "successively introduced" in antecedent lives. The sequence of events resultant on its "unwinding" would be the destiny of the person—a destiny determined, necessarily, by past action. This concept gives a new and more eloquent meaning to the phrase "Character is destiny." If we carry our thought no further, we are plunged into the slough of determinism—sheer fatality. But in each reincarnation, however pre-determined every act and event, their reaction upon consciousness remains a matter of determination—is therefore *self*-determined. We may not control the event, but our acceptance of it we may control. Moreover, each "unwinding" of the karmic coil takes place in a new environment, in a world more highly organized by reason of the play upon it of the collective consciousness of mankind. Though the same individual again and again intersects the stream of mundane experience, it is an evolving ego and an augmenting stream. Therefore each life of a given series forms a different, a more intricate, and a more amazing pattern: in each the thread is drawn from nearer the central energy, which is divine, and so shows forth more of the coiled power within the soul.

X

GENIUS

IMMANENCE

THE greatest largess to the mind which higher thought brings is the conviction of a transcendent existence. Though we do not know the nature of this existence, except obscurely, we are assured of its reality and of its immanence, through a growing sense that all that happens to us is simply our relation to it.

In our ant-like efforts to attain to some idea of the nature of this transcendent reality, let us next avail ourselves of the help afforded by the artist and the man of genius, too troubled by the flesh for perfect clarity of vision, too dominated by the spirit not to attempt to render or record those glimpses of *the world-order* now and then vouchsafed. For the genius stands midway between man and Beyond-man: in Nietzche's phrase, "Man is a bridge and not a goal."

Of all the writers on the subject of genius, Schopenhauer is the most illuminating, perhaps because he so suffered from it. According to him, the essence of genius lies in the perfection and energy of its *perceptions.* Schopenhauer says, "He who is endowed with talent thinks more quickly and more correctly than others; but the genius beholds another world from them all, although only because he has a more profound perception of the world which lies before them also, in that it presents itself in his mind more objectively, and consequently in greater purity and distinctness." This more profound perception arises from his detachment: his intellect has to a certain extent freed itself from the service of his will, and leads an independent life. So long as the intellect is in the service of the will that which has no relation to the will does not exist for the intellect; but along with this partial severance of the two there comes a new power of perception, synthetic in its nature, a complex of relationships not reproducible in *linear* thought, for the mind is oriented simultaneously in *many different directions.* Of this order of perception the well-known case of Mozart is a classic example. He is reported to have said of his manner of composing, "I can see the whole of it in my mind at a single glance . . . in which way I do not hear it in my imagination at all as succession—the way it comes later—but all at once, as it were. It is a rare feast! all the inventing and making goes on in me as in a beautiful strong dream."

TIMELESSNESS

The inspirations of genius come from a failure of attention to life, which, all paradoxically, brings vision—the power to see life steadily and *see it whole*. Consciousness, unconditioned by time, "in a beautiful strong dream," awakens to the perception of a world that is timeless. It brings thence some immortelle whose power of survival establishes the authenticity of the inspiration. However local and personal any masterpiece may be, it escapes by some potent magic all geographical and temporal categories, and appears always new-born from a sphere in which such categories do not exist.

No writer was more of his period than Shakespeare, yet how contemporary he seems to each succeeding generation. Leonardo, in a perfect portrait, showed forth the face of a subtle, sensuous, and mocking spirit, against a background of wild rocks. It represents not alone the soul-phase of the later Renaissance, but of every individual and of every civilization which on life's dangerous and orgiastic substratum has reared a mere garden of delight. Living hearts throb to the music penned by the dead hand of Mozart and of Beethoven; the clownings of Aristophanes arouse laughter in our music halls; Euripides is as subtle and world-weary as any modern; the philosophies of Parmenides and

Heraclitus are recrudescent in that of Bergson; and Plato discusses higher space under a different name.

BEYOND GOOD AND EVIL: BEAUTY

The second characteristic of works of genius is their indifference to all man-made moral standards. They are beyond all that goes by the name of good and evil, in that the two are used indifferently for the furtherance of a purely æsthetic end. The Beyond-man discovers beauty in the abyss, and ugliness in mere worldly rectitude. Leonardo painted the Medusa head, with its charnel pallor and its crown of writhing snakes, no less lovingly than the sweet-tender face of the Christ of the *Cenacolo,* and the beauty is not less, though of an opposite sort. Shakespeare's most profound sayings and most magical poetry are as often as not put in the mouths of his villains and his clowns. To genius, pain is purgation; ugliness—beauty in disturbance. It injects the acid of irony into success, and distils the attar of felicity from failure. It teaches that the blows of fate are aimed not at us, but at our fetters; that death is swallowed up in victory; that the Hound of Heaven is none other than the Love of God.

Though genius rebels at our moralities, it always submits itself to beauty. Emerson says, "Goethe and Carlyle, and perhaps Novalis, have an undisguised dislike or contempt for common virtue standing on common principles. Meantime they are dear lovers, steadfast maintainers of the pure, ideal morality. But they worship it as the highest beauty, their love is artistic." And so it is throughout the whole hierarchy of men of genius. "Beauty is Truth: Truth—Beauty," is the motto which guides their far-faring feet, as they lead us wheresoever they will. With Victor Hugo we follow undisgusted through the sewers of old Paris: his sense of beauty disinfects them for us. With Balzac and Tolstoy we gaze unrevolted upon the nethermost depths of human depravity, discerning moral beauty even there; while with Virgil, Dante and Milton, we walk unscathed in Hell itself. The *terribilita* of Michaelangelo, the chaos and anarchy of Shakespeare at his greatest, as in Lear—these find expression in perfect rhythms, so potent that we recognize them as proceeding from a supernal beauty, the beauty of that soul "from which also cometh the life of man and of beast, and of the birds of the air and of the fishes of the sea."

THE DÆMONIC

"Unknown,—albeit lying near,—
To men the path to the Dæmon sphere."

But to men of genius—"Minions of the Morning Star"—the path is not unknown, and for this reason the dæmonic element constantly shows itself in their works and in their lives. Dante, Cellini, Goethe—three men as unlike in the nature of their several gifts and in their temperaments as could easily be named together—are drawn to a common likeness through the dæmonic gleam which plays and hovers over them at times. With William Blake it was a flame that wrapped him round. Today no one knows how Brunelleschi was able to construct his great dome without centering, nor how Michaelangelo could limn his terrible figures on the wet plaster of the Sistine vault with such extraordinary swiftness and skill; but we have their testimony that they invoked and received divine aid. Shakespeare, the master-magician, is silent on this point of supernatural assistance—as on all points—except as his plays speak for him; but how eloquently they speak! *The Tempest* is made up of the dæmonic; the murky tragedy of *Macbeth* unfolds under the guidance of incarnate forces of evil which drive the hero to his doom and final deliverance in death: Hamlet sees and communes with the ghost of his father; in short, the supernatural is as much a part of these plays as salt is part of the ocean. If from any masterpiece we could abstract everything not strictly rational—every element of wonder, mystery, and enchantment—it would be like taking all of the unknown quantities out of an equation: there would be nothing left to solve. The mind of genius is a wireless station attuned to the vibrations from the dæmonic sphere; the works of genius fascinate and delight us largely for this reason: we also respond to these vibrations and are demonologists in our secret hearts.

For the interest which we take in genius has its root in the interest which we take in ourselves. Genius but utters experiences common to us all, records perceptions of a world-order which we too have glimpsed. Love, hope, pain, sorrow, disappointment, often effect that momentary purgation which enables consciousness to function independently of the tyrant will. These hours have for us a noetic value—"some veil did fall" revealing visions remembered even unto the hour of death.

DEATH

That "failure of attention to life" which begets inspiration in the man of genius comes indeed daily to every one, but without his being able to profit by it. For what is sleep but a failure of attention to life—so complete a failure that memory brings back nothing save that little caught in the net of dreams, yet even this little is so charged with creative energy as to give rise to the saying that every man is a genius in his dreams.

Death also is a failure of attention to life, the greatest that we know, and poorest therefore in salvage from supernatural realms. Nevertheless reports of persons who have narrowly escaped death give evidence at least that to those emancipated by death, life, viewed from some higher region of space, is perceived as a unity. When a man is brought face to face with death, the events of life pass before the mind's eye in an instant, and he comes from such an experience not only with deeper insight into himself, but into the meaning and purpose of life also. The faces of the dead, those parchments whereon are written the last testament of the departed spirit, bear an expression of solemn peace, sometimes of joy, sometimes of wonder: terror and agony are seldom written there, save when the fatal change comes in some painful or unnatural way.

THE PLAY OF BRAHM

Inspiration, dreams, visions at the moment of death—these things we say are *irrational,* and so in a sense they are. Bergson has compared the play of reason upon phenomena to the action of a cinematograph machine which reproduces the effect of motion by flashing upon the screen a correlated series of *fixed* images. In like manner the reason dissects the flux of life and presents it to consciousness part by part, but never as a whole. In supernormal states however we may assume that with the breakdown of some barrier life flows in like a tidal wave, paralyzing the reason, and therefore presenting itself in an irrational manner to consciousness. Were reason equal to the strain put upon it under these circumstances, in what light might the phantasmagoria of human life appear? Might it not be perceived as a representation, merely, of a supernal world, higher-dimensional in relation to our own? Just as a moving picture shows us the round and living bodies of men and women as flat images on a plane, enacting there some mimic drama, so on the three-dimensional screen of the world men and women engaged in unfolding the drama of personal life may be but the images of souls enacting, on higher planes of being, the drama of their own salvation. The reluctance of the American aborigine to be photographed is said to have been due to his belief that something of his personality, his human potency, went into the image, leaving him by so much the poorer from that time forth. Suppose such were indeed the case: that the flat-man on the moving picture screen leads his little life of thought and emotion, related to the mental and emotional life of the living original as the body is related to its photographic counterpart. In similar manner the potencies of the higher self, the dweller in higher spaces, may flow into and express themselves in and through us. We may be images in a world of images; our thoughts shadows of archetypal ideas, our acts a shadow-play upon the luminous screen of material existence, revealing there, however imperfectly, the moods and movements of a higher self in a higher space.

The saying, "All the world's a stage," may be true in a sense Shakespeare never intended. It formulates, in effect, the oldest of all philosophical doctrines, that contained in the *Upanishads*, of Brahma the Enjoyer, who takes the form of a mechanically perfect universe in order to read his own law with eyes of his own creation. "He thought: 'Shall I send forth worlds?' He sent forth these worlds." To the question, "What worlds?" the Higher Space Hypothesis makes answer, "Dimensional systems, from lowest to highest, each one a *representation* of the one next above, where it stands *dramatized*, as it were, and the whole a representation or dramatization of *consciousness.* This is the *play of Brahm:* endlessly to dissect and dissever, then to rediscover and reunite, as does the geometrician who discovers every ellipse, parabola, and hyperbola in the cone where all inhere.

The particular act of the drama of unfolding consciousness upon which the curtain is now upfurled is that wherein we discover the world to be indeed a stage, a playground for forces masquerading as forms: "They have their exits and their entrances," or, as expressed in the *Upanishads,* "All that goes hence (dies on earth) heaven consumes it all; and all that goes thence (returns from heaven to a new life) the earth consumes it all."

XI

THE GIFT OF FREEDOM

CONCEPT AND CONDUCT

A SURGEON once remarked to the author that among his professional associates he had noticed an increasing awareness of the invisible. This he claimed was manifest in the fact that the young men educated since the rise of bacteriological science were more punctilious in the matter of extreme personal cleanliness and the sterilization of their instruments than the older and often more accomplished surgeons whose habits in these matters had been formed before the general sense of an *invisible* menace had become acute.

This anecdote well illustrates the unconscious reaction of new concepts upon conduct. Preoccupation with the problems of space hyper-dimensionality cannot fail to produce profound changes in our ethical outlook upon life and in our attitude towards our fellow beings. The nature of these changes it is not difficult to forecast.

Although higher-space thought makes painfully clear our limitations, it nevertheless leads to the perception that these very limitations are inhibited powers. In this way it supplies us with a workable method whereby we may enter that transcendental world of which we glimpse so many vistas. This method consists in first becoming aware of a limitation, and then in forcing ourselves to dramatize the experience that would be ours if the limitation did not affect us. We then discover in ourselves a power for transcending the limitation, and presently we come to live in the new mode as easily as in the old. Thought, conscious of its own limitations, leads to the New Freedom. "Become what thou art!" is the maxim engraved upon the lintel of this New Temple of Initiation.

SELFLESSNESS

Higher-space speculation is an education in *selflessness,* for it demands the elimination of what Hinton calls *self-elements* of observation. The diurnal motion of the sun is an example of a self-element: it has nothing to do with the sun but everything to do with the observer. The Ptolemaic system founded on this illusion tyrannized over the human mind for centuries, but who knows of how many other illusions we continue to be victims—for the worst of a self-element is that its presence is never dreamed of until it is done away with. The Theory of Relativity presents us with an effort to get rid of the self-element in regard to space and time. A self-centered man cannot do full justice to this theory: it requires of the mind a certain detachment, and the idea becomes clear in proportion as this detachment, this selflessness, is attained.

So while it would be too much to claim that higher thought makes men unselfish, it at least cracks the hard shell in which their selfishness abides. If a man disciplines himself to abdicate his personal point of view in thinking about the world he lives in, it makes easier a similar attitude in relation to his fellow men.

HUMILITY

One of the earliest effects of selfless thought is the exorcism of all arrogance. The effort to dramatize the relation of an earthworm to its environment makes us recognize that its predicament is our own, different only in degree. We are exercising ourselves in humility and meekness, but of a sort leading to a mastery that may well make the meek the inheritors of the earth. Hinton was himself so meek a man that his desire did not rise to the height of expecting or looking for the beautiful or the good: he simply asked for something to know. He despaired of knowing anything definitely and certainly except arrangements in space. We have his testimony as to how abundantly this hunger and thirst after that right knowledge which is righteousness was gratified. "All I want to do," he says, "is to make this humble beginning of knowledge and show how inevitably, by devotion to it, it leads to marvelous and far-distant truths, and how, by strange paths, it leads directly into the presence of some of the highest conceptions which great minds have given us."

Here speaks the blessed man referred to by the psalmist, "Whose delight is in the law of the Lord, and in His law doth he meditate day and night." Abandoning a vain search after abstractions, and applying his simple formula to life, Hinton found that it enabled him to express the faith in his heart in terms comformable to reason; that it led back to and illumined the teachings of every spiritual instructor and inspirer of mankind.

SOLIDARITY

That we are all members of one body, branches of one vine, is a matter of faith and of feeling; but with the first use of the weapon of higher thought the paradox of the one and the many is capable of so clear and simple a resolution that the sublime idea of human solidarity is brought down from the nebulous heaven of the mystic to the earth of every day life. To our ordinary space-thought, men are isolated, distinct, each "an infinitely repellent particle," but we conceive of space too narrowly. The broader view admits the idea that men are related by reason of a superior union, that their isolation is but an affair of limited consciousness. Applying this concept to conduct, we come to discern a literal truth in the words of the Master, "He who hath done it unto the least of these my children, hath done it unto me," and "Where two or three are gathered together in my name." If we conceive of each individual as a "slice" or cross-section of a higher being, each fragment isolated by an inhibition of consciousness which it is moment by moment engaged in transcending, the sacrifice of the Logos takes on a new meaning. This disseverance into millions of human beings is that each may realize God in himself. Conceiving of humanity as God's broken body, we are driven to make peace among its members, and by realization we become the Children of God.

LIVE OPENLY

"Blessed are the meek." "Blessed are they that hunger and thirst after righteousness." "Blessed are the peacemakers." These beatitudes are seen to have a direct relation to higher-space thought; nor would it be impossible to trace a relation between this thought and the other beatitudes also, but it will suffice simply to note the fact that the central and essential teaching of the Sermon on the Mount,

"Let your light shine before men" is implicit in the conviction of every one who thinks on higher space: he must *live openly.* By continual dwelling upon the predicament of the flat-man, naked, as it were, to observation from an eye which looks down upon his plane, we come to realize our own exposure. In that large world all that we think, or do, or imagine, lies open, palpable; there is no such thing as secrecy. Imbued with this idea, we begin to live openly because we must; but soon we come to do so because we desire it. In making toward one another our limited lives open and manifest, we treat each other in the service of truth as though we were all members of that higher world. We imitate in our world our true existence in a higher world, and so help to establish heavenly conditions upon earth.

NON-RESISTANCE TO EVIL

The problem of ugliness and evil would seem at first thought to be totally unrelated to the subject of space hyper-dimensionality, but there is at least a symbolical relation. This was suggested to the author by the endeavor of two friends, whose interests were preëminently mathematical, to discover what certain four-dimensional figures would look like in three-dimensional space. They found that in a great number of cases these cross-sections, when thus isolated, revealed little of the symmetry and beauty of their higher-dimensional prototypes. It is clear that a beautiful form of our world, traversing a plane, would show nothing of its beauty to the plane-man, who lacked the power of perceiving it entire; for the sense of beauty is largely a matter of coördination. We give the names of evil, chance, fate, ugliness, to those aspects of life and of the world that we fail to perceive in their true relations, in regard to which our power of correlation breaks down. Yet we often find that in the light of fuller knowledge or subsequent experience the fortune which seemed evil was really good fortune in the making, that the chance act or encounter was too momentous in its consequences to be regarded as other than ordained.

The self-element plays a large part in our idea of good and evil, ugliness and beauty: "All things are as they seem to all." Desire of her will make any woman beautiful, and fear will exercise an absolute inhibition upon the æsthetic sense. As we recede in time from events, they more and more emancipate themselves from the tyranny of our personal prejudices and predilections, and we are able to perceive them with greater clarity, more as they appear from the standpoint of higher time and higher space. "Old, unhappy, far-off things, and battles long ago" lose their poignancy of pain and take on the poignancy of beauty. The memory of suffering endured is often the last thing from which we would be parted, while humdrum happiness we are quite willing to forget. Because we realize completely only in retrospect, it may well be that the present exists chiefly for the sake of the future. Then let the days come with veiled faces, accept their gifts whose value we are so little able to appraise! There is a profound and practical truth in Christ's saying, "Resist not evil." Honor this truth by use, and welcome destiny in however sinister a guise!

THE IMMANENT DIVINE

In the fact of the limited nature of our space perceptions is found a connecting link between materialism and idealism. For, passing deeper and deeper in our observation of the material world, that which we at first felt as real passes away to become but the outward sign of a reality infinitely greater, of which our realities are appearances only, and we become convinced of the existence of *an immanent divine:* "In Him we live and move and have our being." Our space is but a limitation of infinite "room to move about": *"In my Father's house are many mansions."* Our time is but a limitation of infinite duration: *"Before Abraham was, I am."* Our sense of space is the consciousness that we abide in Him; our sense of time is the consciousness that He abides in us. Both are modes of apprehension of divinity—growing, expanding modes.

In conceiving of a space of more than three dimensions we prove that our relation to God is not static, but dynamic. Christ said to the man who was sick of the palsy, "Rise, take up thy bed and walk." The narrow concept of three-dimensional space is a bed in which the human mind has lain so long as to become at last inanimate. The divine voice calls to us again to demonstrate that we are alive. Thinking in terms of the higher we issue from the tomb of materialism into the sunlight of that sane and life-giving idealism which is Christ's.

COSIMO CLASSICS

CPSIA information can be obtained at www.ICGtesting.com
Printed in the USA
LVOW06s1521270514

387445LV00001B/253/P